T0219940

Lecture Notes in Artificial Intelligence 10872

Subseries of Lecture Notes in Computer Science

More information about this series at http://www.springer.com/series/1244

Peter Chapman · Dominik Endres
Nathalie Pernelle (Eds.)

Graph-Based Representation and Reasoning

23rd International Conference
on Conceptual Structures, ICCS 2018
Edinburgh, UK, June 20–22, 2018
Proceedings

 Springer

Editors
Peter Chapman ⓘ
Edinburgh Napier University
Edinburgh
UK

Nathalie Pernelle
University of Paris-Sud
Orsay
France

Dominik Endres
Philipps University of Marburg
Marburg
Germany

ISSN 0302-9743   ISSN 1611-3349   (electronic)
Lecture Notes in Artificial Intelligence
ISBN 978-3-319-91378-0   ISBN 978-3-319-91379-7   (eBook)
https://doi.org/10.1007/978-3-319-91379-7

Library of Congress Control Number: 2018944283

LNCS Sublibrary: SL7 – Artificial Intelligence

Printed on acid-free paper

This Springer imprint is published by the registered company Springer International Publishing AG
part of Springer Nature
The registered company address is: Gewerbestrasse 11, 6330 Cham, Switzerland

Preface

The 23rd International Conference on Conceptual Structures (ICCS 2018) took place at Edinburgh Napier University during June 20–22, 2018. Since its inception in 1993, the research presented at ICCS has focussed on representation of and reasoning with conceptual structures in a variety of contexts. For example, at the 19th ICCS the context was knowledge discovery in social network data. Two years later, at the 20th ICCS, the conceptual structures of STEM research and education were investigated. This year, the theme of ICCS was synergies between human and machines. Interest in this topic was indicated by a substantial number of contributions addressing this theme, as well as the keynote "Modular Ontologies as a Bridge Between Human Conceptualization and Data," delivered by Prof. Pascal Hitzler, director of the Data Science and Data Semantics (DaSe) Laboratory, Wright State University at Dayton, Ohio, USA. His research interests cover an impressive range from theory to application, including mathematical foundations of computer science and artificial intelligence, Semantic Web, linked and big data, ontologies, neuro-symbolic integration, knowledge representation, and of course, conceptual structures. He is also an Endowed NCR Distinguished Professor, so we were very glad when he agreed to present his ideas at ICCS.

This year, ICCS was co-located with the 10th International Conference on the Theory and Application of Diagrams (Diagrams 2018). To promote interactions between the researchers of both communities further, the conferences shared a keynote by Keith Stenning on "Diagrams and Non-Monotonic Logic: What Is the Cognitive Relation?"

Keith Stenning was in an excellent position for delivering a joint keynote. As a psychologist who researches the evolution of human cognitive capacities in general, and reasoning with diagrams as opposed to purely linguistic presentations in particular, his research foci resonate well with this year's ICCS theme and Diagrams.

Of the regular submissions received, we were able to accept ten as full papers (56%), three as short papers, and two as posters. The decisions for inclusion in this volume were based on reviews from at least three, often four, and in one case five, expert members of the Program Committee, after the authors had a chance to submit a rebuttal to the initial reviews. We believe this procedure ensured that only high-quality contributions were presented at the conference. We would like to thank the Program Committee members and the additional reviewers for their hard work. Without their substantial voluntary contribution, it would have been impossible to ensure a high-quality conference program.

For the purpose of presentation at the conference, we divided the accepted submissions into five sessions: graph- and concept-based inference, human–machine interaction, modeling human cognition, graph visualization, and a poster session. Such a division is necessarily a compromise, but we think it captured the main themes of the respective contributions.

There are, of course, many people to whom we are indebted for their considerable assistance in making ICCS 2018 a success. Our institutions, Edinburgh Napier University, the University of Marburg, and the University of Paris, also provided support for our participation, for which we are grateful. Lastly, we thank the ICCS Steering Committee for their continual support, advice, and encouragement.

June 2018   Peter Chapman
                                                              Dominik Endres
                                                            Nathalie Pernelle

Organization

Program Committee

Simon Andrews	Sheffield Hallam University, UK
Moulin Bernard	Laval University, Canada
Peggy Cellier	IRISA/INSA Rennes, France
Peter Chapman	Edinburgh Napier University, UK
Dan Corbett	Optimodal Technologies, USA
Olivier Corby	Inria, France
Madalina Croitoru	LIRMM, University of Montpellier II, France
Licong Cui	University of Kentucky, USA
Juliette Dibie-Barthélemy	AgroParisTech, France
Pavlin Dobrev	Bosch Software Innovations, Germany
Florent Domenach	Akita International University, Japan
Dominik Endres	University of Marburg, Germany
Catherine Faron Zucker	Université Nice Sophia Antipolis, France
Ollivier Haemmerlé	IRIT, University of Toulouse le Mirail, France
Jan Hladik	DHBW Stuttgart, Germany
John Howse	University of Brighton, UK
Dmitry Ignatov	National Research University Higher School of Economics, Russia
Mateja Jamnik	University of Cambridge, UK
Adil Kabbaj	INSEA, USA
Mary Keeler	VivoMind, Inc., USA
Steffen Lohmann	Fraunhofer, Germany
Natalia Loukachevitch	Research Computing Center of Moscow State University, Russia
Pierre Marquis	CRIL, University of Artois and CNRS, France
Franck Michel	Université Côte d'Azur, CNRS, I3S, France
Sergei Obiedkov	National Research University Higher School of Economics, Russia
Yoshiaki Okubo	Hokkaido University, Japan
Nathalie Pernelle	LRI-Université Paris Sud, France
Heather D. Pfeiffer	Akamai Physics, Inc., USA
Simon Polovina	Sheffield Hallam University, UK
Uta Priss	Ostfalia University, Germany
Sebastian Rudolph	TU Dresden, Germany
Eric Salvat	IMERIR, France
Fatiha Saïs	LRI (CNRS and Université Paris-Sud 11) and Inria Saclay, France
Iain Stalker	University of Manchester, UK

Gem Stapleton	University of Brighton, UK
Michaël Thomazo	Inria, Université Paris Saclay and LIX, Ecole Polytechnique, Université Paris Saclay, France
Serena Villata	CNRS - Laboratoire d'Informatique, Signaux et Systèmes de Sophia-Antipolis, France
Martin Watmough	Sheffield Hallam University, UK

Additional Reviewers

Abeysinghe, Rashmie
Delobelle, Jérôme
Groz, Benoit
Liu, Jinpeng
Schwind, Nicolas
Sim, Kevin
Zheng, Fengbo

Diagrams and Nonmonotonic Logic:
What is the Cognitive Relation?
(Abstract)

Keith Stenning

University of Edinburgh

Diagrams originally caught my attention while I was, as usual, researching natural language discourse. They offered a counterpoint to language. Being told that using logical analysis of natural language discourse means the psychology is limited to linguistic reasoning had grown tedious. My first love had been the idea that there was an alternative logic for the construction of models for narratives. The counterpoint of diagrams reinforced the idea that interpretation had to be central. But in natural language narrative processing, the involvement of processing is far more all encompassing. Unlike the situation in classical logic where interpretation is a starter for the real meal, in discourse processing, interpretation is the main course. The problem had been in the early years [2], that classical logic was obviously the wrong logic, but there wasn't as yet a suitable alternative. Nonmonotonic logics had been invented [1] but it wasn't easy to see how they could be applied. Although they had been invented to make 'ordinary' reasoning with general knowledge easy, it had turned out that their tractability was even worse than classical logic, and here it wasn't clear that they could be fitted to small problems, because they could immediately explode by demanding the retrieval of just about any piece of human knowledge to connect the first two sentences of a discourse.

So the real purpose of this talk to this audience is to explore the similarities between nonmonotonic logics and diagrams. Logic Programming (LP), the nonmonotonic logic that grew out of PROLOG, is 'weak' in a number of ways. Its birth gives the most general hint. How did PROLOG arise? From the gleam in the eye that logic ought to be usable for programming computers at a much higher level than the then available imperative languages, combined with the ghastly realisation that classical logic was fundamentally intractable. Much of our *un*derstanding of computation came from the study of the undecidability of classical logic. So LP is based on the 'Horn-clause fragment' of classical logic, which was the most promising fragment that is tractable. So weakness was the fundamental desideratum (coupled of course with just enough expressiveness for a programming language).

References

1. McCarthy, J.: Circumscriptiona form of non-monotonic reasoning. In: Readings in Artificial Intelligence, pp. 466–472. Elsevier (1981)
2. Stenning, K.: Anaphora as an Approach to Pragmatics. MIT Press (1978)

Contents

Invited Keynote

Modular Ontologies as a Bridge Between Human Conceptualization
and Data   3
Pascal Hitzler and Cogan Shimizu

Graph- and Concept-Based Inference

Using Analogical Complexes to Improve Human Reasoning and Decision
Making in Electronic Health Record Systems   9
Christian Săcărea, Diana Şotropa, and Diana Troancă

An Efficient Approximation of Concept Stability
Using Low-Discrepancy Sampling.   24
Mohamed-Hamza Ibrahim and Rokia Missaoui

Lifted Most Probable Explanation   39
Tanya Braun and Ralf Möller

Lifted Dynamic Junction Tree Algorithm   55
Marcel Gehrke, Tanya Braun, and Ralf Möller

Computer Human Interaction and Human Cognition

Defining Argumentation Attacks in Practice: An Experiment
in Food Packaging Consumer Expectations   73
Bruno Yun, Rallou Thomopoulos, Pierre Bisquert,
and Madalina Croitoru

Empirically Evaluating the Similarity Model of Geist,
Lengnink and Wille.   88
Moritz Schubert and Dominik Endres

Combining and Contrasting Formal Concept Analysis and APOS Theory. . . .   96
Uta Priss

Musical Descriptions Based on Formal Concept Analysis
and Mathematical Morphology 105
Carlos Agon, Moreno Andreatta, Jamal Atif, Isabelle Bloch,
and Pierre Mascarade

Towards Collaborative Conceptual Exploration 120
Tom Hanika and Jens Zumbrägel

Graph Visualization

A Visual Analytics Technique for Exploring Gene Expression
in the Developing Mouse Embryo............................. 137
    Simon Andrews and Kenneth McLeod

Exploring Heterogeneous Sequential Data on River Networks
with Relational Concept Analysis............................. 152
    Cristina Nica, Agnès Braud, and Florence Le Ber

Conceptual Graphs Based Modeling of Semi-structured Data............ 167
    Viorica Varga, Christian Săcărea, and Andrea Eva Molnar

Using Conceptual Structures in Enterprise Architecture to Develop
a New Way of Thinking and Working for Organisations............... 176
    Simon Polovina and Mark von Rosing

Posters

Visualizing Conceptual Structures Using FCA Tools Bundle.......... 193
    Levente Lorand Kis, Christian Săcărea, and Diana-Florina Şotropa

Node-Link Diagrams as Lenses for Organizational Knowledge Sharing
on a Social Business Platform............................... 197
    Anne-Roos Bakker and Leonie Bosveld-de Smet

Author Index... 201

Invited Keynote

Modular Ontologies as a Bridge Between Human Conceptualization and Data

Pascal Hitzler[✉] and Cogan Shimizu

Data Semantics Laboratory, Wright State University, Dayton, OH, USA
pascal.hitzler@wright.edu

Abstract. Ontologies can be viewed as the middle layer between pure human conceptualization and machine readability. However, they have not lived up to their promises so far. Most ontologies are too tailored to specific data and use-cases. By making sometimes strong, or sometimes too weak, ontological commitments, many existing ontologies do not adequatly reflect human conceptualizations. As a result, sharing and reuse of ontologies is greatly inhibited. In order to more effectively preserve this notion of human conceptualization, an ontology should be designed with modularity and extensibility in mind. A modular ontology thus may act as a bridge between human conceptualization and data.

1 The Case for Modular Ontologies

The Internet is the single largest repository of knowledge to have ever existed and continues to grow every second. The amount of data continuously generated by both humans and machines defies comprehension: from second-by-second meteorological data gathered by sensors to academic articles written by scientists to communications on social media networks to collaborative articles on Wikipedia. How can we represent and link these disparate forms of data together in order to generate an understandable gestalt? We would require a way to organize acquired data such that some critical part of the human conceptualization of each piece is preserved.

Ontologies, as "explicit specifications of conceptualizations," seem like a natural fit for the role [2]. With the explosive growth of the Semantic Web in the last decade, it would seem that they have seen no small success for that purpose. Ontologies offer a human accessible organization of immense amounts of data and act as a vehicle for the sharing and reuse of knowledge.

Unfortunately, published ontologies have often not lived up to these promises. Large, monolithic ontologies, designed with very strong – or very weak – ontological commitments are very difficult to reuse across the same domain, let alone different domains. Strong ontological commitments lead to overspecification, to ontologies essentially being only fit for the singular purpose for which they were originally designed. Weak ones lead to ambiguity of the model, sometimes to the extent that is hard to grasp what is actually being modeled.

© Springer International Publishing AG, part of Springer Nature 2018
P. Chapman et al. (Eds.): ICCS 2018, LNAI 10872, pp. 3–6, 2018.
https://doi.org/10.1007/978-3-319-91379-7_1

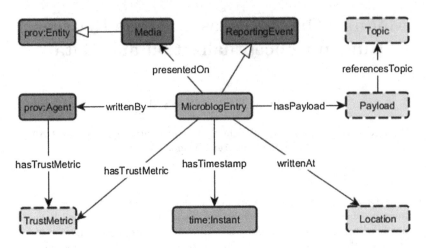

Fig. 1. This is a graphical view of the MicroblogEntry ODP [12]. Yellow boxes indicate datatypes, light blue boxes with dashed borders indicate external patterns. Purple is used for external classes belonging to PROV-O [1]. Green is used for external classes belonging to [8]. White arrowheads represent the owl:SubclassOf relation. (Color figure online)

We posit that one effective way to obtain ontologies which are easier to reuse, is to build them in a modular fashion. A sufficiently modularized ontology [9] is designed such that individual users can easily adapt an ontology to their use cases, while maintaining integration and relationships with other versions of the ontology. A modular ontology is constructed by piecing together so-called ontology modules. Ontology modules are created by adapting Ontology Design Patterns to the domain and use-case [4,5].

2   Ontology Design Patterns

In general, patterns are invariances that may be observed over different media (e.g. data or processes). Ontology Design Patterns (ODP) are the recognition of *conceptual* patterns that occur across different domains. Modeling with ODPs has established itself as an ontology engineering paradigm [5].

The Semantic Trajectory ODP [7] is a classic example of a recurring pattern. However, patterns are designed to be sufficiently general as to apply to many different cases as possible. Thus, it is necessary to create a module from them by adapting the pattern to the specific domain and use-case in mind. The Semantic Trajectory ODP has been successfully modularized a number of times; two prominent examples are the CruiseTrajectory ODP [11] and the SpatiotemporalExtent ODP [10]. For a thorough tutorial on creating modules out of patterns, see [9].

As another example, Fig. 1 shows a graphical representation of the MicroblogEntry ODP [12]. This ODP clearly demonstrates pattern reuse and how adequate ontological commitment eases of modularization. This ODP was engineered to leverage as much existing work as possible. For example, the Media and ReportingEvent concepts are defined in [8]. The concepts Entity and Agent come from the popular PROV Ontology that express provenance data [1]. Further, this ODP avoids overly strong ontological commitments, allowing it to be easily modularized to represent specific microblogs (e.g. Twitter vs. Facebook vs. Instagram).

3   The Future of Modular Ontology Engineering

The promise of modular ontologies is still being realized. There are yet open questions concerning ontology design patterns, their usage, and the surrounding supporting tools and infrastructure. For a more thorough examination of these questions and challenges, see [3]. That is not to say that there are no efforts underway; here, we briefly identify some of these ongoing efforts.

Perhaps the most fundamental purpose of the Semantic Web is to enable the sharing and reuse of knowledge. Certainly, a pattern is knowledge in and of itself. Thus, it is only reasonable that there needs to be a way to enable the sharing and reuse of patterns, as well. To do so, we are working towards the development of a "smart," central repository. Such a repository would be initially populated with a critical mass of fundamental ODPs. That is, a collection of ODPs with sufficient breadth and generalization such that their combination covers any complex conceptualization.

In addition, these patterns will be annotated in a systematic and rigorous way. Answering questions such as

- How do patterns interact with each other?
- Do they import other patterns?
- Which pattern did this module reuse as a template?

Recently, [6] introduced the Ontology Design Pattern Representation Language (OPLa) as a way to address those questions, and others. The smart repository would use these OPLa annotations in order to inform an ontology engineer on available patterns, especially those related to their domain and use-cases.

Between the central repository and OPLa, the next step will be to create a graphical interface for the assembly and modularization of ontologies. This will be a combination of different visualization strategies and a plug-and-play system for ODPs.

And finally, as we learn how to most help humans create ontologies, can we attempt to also automate these processes? That is, automatically create an ontology from a dataset and present it as a "first draft" to the ontology engineer for editing?

Acknowledgement. Cogan Shimizu acknowledges support by the Dayton Area Graduate Studies Institute (DAGSI).

References

1. Groth. P., Moreau, V. (eds.): PROV-Overview: An Overview of the PROV Family of Documents. W3C Working Group Note 30 April 2013 (2013)
2. Gruber, T.R.: A translation approach to portable ontology specifications. Knowl. Acquis. **5**(2), 199–220 (1993)
3. Hammar, K., Blomqvist, E., Carral, D., van Erp, M., Fokkens, A., Gangemi, A., van Hage, W.R., Hitzler, P., Janowicz, K., Karima, N., Krisnadhi, A., Narock, T., Segers, R., Solanki, M., Svátek, V.: Collected research questions concerning ontology design patterns. In: Hitzler, P., et al. (eds.) Ontology Engineering with Ontology Design Patterns - Foundations and Applications. Studies on the Semantic Web, vol. 25, pp. 189–198. IOS Press, Amsterdam (2016)
4. Hammar, K., Hitzler, P., Krisnadhi, A., Nuzzolese, A.G., Solanki, M. (eds.): Advances in Ontology Design and Patterns. Studies on the Semantic Web, vol. 32. IOS Press, Amsterdam (2017)
5. Hitzler, P., Gangemi, A., Janowicz, K., Krisnadhi, A., Presutti, V. (eds.): Ontology Engineering with Ontology Design Patterns - Foundations and Applications. Studies on the Semantic Web, vol. 25. IOS Press, Amsterdam (2016)
6. Hitzler, P., Gangemi, A., Janowicz, K., Krisnadhi, A.A., Presutti, V.: Towards a simple but useful ontology design pattern representation language. In: Blomqvist, E., et al. (eds.) Proceedings of the 8th Workshop on Ontology Design and Patterns (WOP 2017), CEUR Workshop Proceedings, vol. 2043. CEUR-WS.org, Vienna (2017)
7. Hu, Y., Janowicz, K., Carral, D., Scheider, S., Kuhn, W., Berg-Cross, G., Hitzler, P., Dean, M., Kolas, D.: A geo-ontology design pattern for semantic trajectories. In: Tenbrink, T., Stell, J., Galton, A., Wood, Z. (eds.) COSIT 2013. LNCS, vol. 8116, pp. 438–456. Springer, Cham (2013). https://doi.org/10.1007/978-3-319-01790-7_24
8. Kowalczuk, E., Lawrynowicz, A.: The reporting event ODP and its extension to report news events. In: Hammar, K., et al. (eds.) Advances in Ontology Design and Patterns. Studies on the Semantic Web, vol. 32, pp. 105–117. IOS Press, Amsterdam (2017)
9. Krisnadhi, A., Hitzler, P.: Modeling with ontology design patterns: chess games as a worked example. In: Hitzler, P., et al. (eds.) Ontology Engineering with Ontology Design Patterns - Foundations and Applications. Studies on the Semantics Web, vol. 25, pp. 3–21. IOS Press, Amsterdam (2016)
10. Krisnadhi, A., Hitzler, P., Janowicz, K.: A spatiotemporal extent pattern based on semantic trajectories. In: Hammar, K., et al. (eds.) Advances in Ontology Design and Patterns. Studies on the Semantic Web, vol. 32, pp. 47–54. IOS Press/AKA Verlag, Amsterdam/Berlin (2017)
11. Krisnadhi, A., Hu, Y., Janowicz, K., Hitzler, P., Arko, R., Carbotte, S., Chandler, C., Cheatham, M., Fils, D., Finin, T., Ji, P., Jones, M., Karima, N., Lehnert, K., Mickle, A., Narock, T., O'Brien, M., Raymond, L., Shepherd, A., Schildhauer, M., Wiebe, P.: The GeoLink modular oceanography ontology. In: Arenas, M., Corcho, O., Simperl, E., Strohmaier, M., d'Aquin, M., Srinivas, K., Groth, P., Dumontier, M., Heflin, J., Thirunarayan, K., Staab, S. (eds.) ISWC 2015. LNCS, vol. 9367, pp. 301–309. Springer, Cham (2015). https://doi.org/10.1007/978-3-319-25010-6_19
12. Shimizu, C., Cheatham, M.: An ontology design pattern for microblog entries. In: Blomqvist, E., et al. (eds.) Proceedings of the 8th Workshop on Ontology Design and Patterns (WOP 2017), Vienna, Austria, 21 October 2017. CEUR Workshop Proceedings, vol. 2043. CEUR-WS.org (2017)

Graph- and Concept-Based Inference

Using Analogical Complexes to Improve Human Reasoning and Decision Making in Electronic Health Record Systems

Christian Săcărea, Diana Şotropa[✉], and Diana Troancă

Department of Computer Science, Babeş-Bolyai University, Cluj-Napoca, Romania
csacarea@math.ubbcluj.ro, {diana.halita,dianat}@cs.ubbcluj.ro

Abstract. A key ability of human reasoning is analogical reasoning. In this context, an important notion is that of analogical proportions that have been formalized and analyzed in the last decade. A bridging to Formal Concept Analysis (FCA) has been brought by introducing analogical complexes, i.e. formal concepts that share a maximal analogical relation enabling by this analogies between (formal) concepts. Electronic Health Record (EHR) systems are nowadays widespread and used in different scenarios. In this paper we consider the problem of improving EHR systems by using analogical complexes in an FCA based setting. Moreover, we present a study case of analogical complexes in a medical field. We analyze analogical proportions in Electronic Health Record Systems and prove that EHRs can be improved with an FCA grounded analogical reasoning component. This component offers methods for knowledge discovery and knowledge acquisition for medical experts based on patterns revealed by analogies. We also show that combining analogical reasoning with FCA brings a new perspective on the analyzed data that can improve the understanding of the subsequent knowledge structures and offering a valuable support for decision making.

Keywords: Formal Concept Analysis · Analogical complexes
Electronic Health Record Systems

1 Introduction

Conceptual thinking, logical reasoning, identifying analogies and by thus comparing resulting patterns are key features of human intelligence. Conceptual thinking relies on conscious reflexion, discursive argumentation and human communication, where the role of a thinking, arguing and communicating human being is central in the act of processing knowledge and its inherent structures [19]. The mathematization of conceptual knowledge is done by Formal Concept Analysis (FCA). The basic skills and means of FCA are, by this mathematization, the basic means and skills of Conceptual Knowledge Processing. Logical reasoning is done by *judgements*, understood as asserting propositions, and are formally mathematized by *Conceptual Graphs* [18]. Analogical reasoning has

© Springer International Publishing AG, part of Springer Nature 2018
P. Chapman et al. (Eds.): ICCS 2018, LNAI 10872, pp. 9–23, 2018.
https://doi.org/10.1007/978-3-319-91379-7_2

been discussed by Sowa in [16], where an Analogy Engine is described. In that paper, Sowa points out that especially in language understanding, identifying analogies and thus analogical processes provide a greater flexibility than more constrained methods. Moreover, analogy is considered to be a prerequisite of logical reasoning.

The idea to support and enhance the medical system has lasted over time. In order to reach these goals, it is important for clinicians to have the possibility to access patient record information anytime and anywhere. The challenge of providing clinicians of any specialty with an integrated view of the complete healthcare history of the patient has so far proved difficult to meet. Due to this major obstacle, our purpose is to effectively deliver healthcare support for clinicians in order to give them a better perspective over a patient healthcare status.

Digitized health information systems are nowadays widespread as a result of constant development of e-health techniques, advances and methodologies. They are comprising many services and systems, relating healthcare with information technology. Among them, Electronic Health Records (EHR) are one of the frames for patient data communication between healthcare professionals.

Electronic Health Records (EHR) describe the concept of a comprehensive collection of a patient's health and healthcare data. They are electronically maintained information about an individual's lifetime health status and healthcare. It has been intensively studied over the last years whether such electronic systems have improved the medical system and what the benefits and the drawbacks are [5,8,14,17].

The development of Electronic Health Records Systems (EHRs) might be influenced by at least two factors: costs and interoperability. Cost has become a critical factor in healthcare. Technical knowledge and cross-border situations are also an important factor in that influence EHRs improvement. Moreover, the ability of a system to interact with one or more other systems has a huge influence over EHRs.

Traditional models of healthcare services are usually inefficient, being associated with expensive interventions. By using EHR systems one could analyze a variety of data about patient's current health status and medical history (medication, symptoms and diagnostics). Furthermore, Electronic Health Record systems may be used in order to reduce human error and thereby to improve patient safety. One may expect that computerized systems could reduce costs, mistakes or unneeded diagnostic tests.

Analyzing medical data is not a new field of research, due to multiple attempts of interpreting data by using statistical methods, neural networks or decision trees [2]. In our previous research, we have proven, by making use of the effectiveness and the graphical representations of conceptual hierarchies, that FCA and its knowledge processing capabilities are good candidates to solve the knowledge discovery, processing and representation task in EHR systems [4,12]. Similar techniques were used by other researchers, for instance to data mine and interpret patient flows within a healthcare network [6]. Moreover, in our recent research [13], we have managed to offer more insights into how patterns can be extracted from medical data and interpreted by means of FCA and Neo4j, i.e., a graph-based database.

In this paper, we expand our focus and prove that EHR can be enhanced with an analogical reasoning Component. For this, we rely on FCA and analogical proportions, as they have been introduced in [9,10]. Human reasoning is often based on analogical reasoning, but it is also error prone. However, if a computerized system can compute such analogies based on the whole dataset, it could improve the accuracy of the decision process. One of the key notions of analogical reasoning is analogical proportion, which is a statement of the form *a is to b as c is to d*. This statement expresses a similarity between the relation linking *a* to *b* and the relation linking *c* to *d*. Due to different logical and algebraic studies of analogical proportions, reasoning by analogy has been recently improved and formalized. While the use and applications of analogical reasoning are multiple, it was proven, for instance, that analogical reasoning is a powerful tool for classification [1].

On the other hand, FCA is well known for its effective knowledge discovery algorithms and its expressive power, being capable to unify methodologies and find patterns in large datasets. We apply this paradigm in order to obtain in-depth and highly qualitative knowledge representation of medical data. The key notion of FCA is a formal concept, which refers to a maximal subset of elements having a maximal set of properties (attributes) in common.

In the last years, research has suggested FCA and analogical reasoning are two techniques that can be easily combined in the data analysis process, since they both rely on the idea of similarity [9–11]. In this research, we discuss an Analogical Reasoning component grounded on the conceptual landscape paradigm of Wille [3,18,19]. We consider a particular set of medical data to exemplify the FCA based analogies at work. For this, we identify concepts in a weak analogical proportion, i.e., sets of four formal concepts that share a maximal analogical relation. This research proves the effectiveness of FCA together with analogical reasoning as a suitable mechanism through which Electronic Health Record Systems might be improved, offering thus a valuable reasoning and decision making support for medical experts.

The paper is structured as follows. After introducing some preliminaries on EHR systems, FCA and analogical reasoning, we continue with a brief motivation of our medical data analysis. Then, we describe the experiments through which we emphasize how powerful analogical reasoning is by showing what data can be inferred from the results and to what purpose. We end the paper by presenting our conclusion and some future work.

2 Preliminaries

2.1 Formal Concept Analysis

Conceptual Knowledge Processing is an approach to knowledge management which is based on FCA as its underlying mathematical theory. FCA has constantly developed in the last 30 years, rapidly growing from an incipient lattice theory restructuration approach to a mature scientific field with a broad range of applications. Nowadays, FCA comprises not only important theoretical advances about concept lattices, their properties and related structures (like description logics, and pattern structures), but also algorithmics, applications to knowledge discovery and knowledge representation, and several extensions (fuzzy, temporal, relational FCA, etc.).

The basic data type FCA is using is a formal context. Using concept forming operators, a mathematical structure called concept lattice is build. This concept lattice is used as a basis for further communication and analysis. In the following we briefly recall some basic definitions. For more, please consult the standard literature [3].

A *formal context* $\mathbb{K} := (G, M, I)$ consists of two sets G and M and a binary relation I between G and M. The elements of G are called *objects* and the elements of M are called *attributes*. The relation I is called the incidence relation of the formal context, and we sometimes write gIm instead of $(g, m) \in I$. If gIm holds, we say that *the object g has the attribute m*.

Concept forming operators are defined on the power set of G and M, respectively. For $A \subseteq G$, we define $A' := \{m \in M \mid \forall\, g \in A.\ gIm\}$, and for $B \subseteq M$ we have $B' = \{g \in G \mid \forall\, m \in B.\ gIm\}$. These *derivation operators* provide a Galois connection between the power sets of G and M. A *formal concept* is a pair (A, B) with $A \subseteq G$ and $B \subseteq M$ with $A' = B$ and $B' = A$. The main theorem of FCA proves that the set of all concepts of a context $\mathbb{K}$ is a complete lattice and every complete lattice occurs as a concept lattice of a suitable chosen formal context.

Usually data are not binary, so objects might have attributes with some values. A *many-valued* context is a tuple (G, M, W, I), where G, M are sets, $I \subseteq G \times M \times W$ is a ternary relation and for all $g \in G$ and $m \in M$ if $(g, m, w) \in I$ and $(g, m, v) \in I$ then $w = v$, i.e., the value of the object g on the attribute m is uniquely determined.

Many-valued contexts can be binarized using a process called conceptual scaling. By this, we can allocate to every many-valued context a conceptual structure consisting of the many-valued context itself, a set of scales and a conceptual hierarchy.

The conceptual structures encoded in medical data can be extracted using FCA and they can be used as a basis for further communication or knowledge acquisition. Using FCA to construct conceptual maps for medical behavior, the obtained lattice structures of these conceptual maps reflect the characteristics of the medical environment.

2.2   Proportional Analogies for Formal Concepts

Through analogical reasoning one may draw plausible conclusions by exploiting parallels between situations. A key pattern which is associated with the idea of analogical reasoning is the notion of analogical proportion (AP), i.e., a statement between two pairs (A, B) and (C, D) of the form A *is to* B *as* C *is to* D where all elements A, B, C, D are in the same category. This relation express that what is common to A and B is also common to C and D, and what is different between A and B is also different between C and D. There are numerous examples of such statements in our everyday life, i.e. "Paris is to France as Berlin is to Germany", "30 is for 60 what 25 is to 50". Due to the fact that this relation is involving both similarities and dissimilarities between four objects, an analogical proportion is able to model any complex association.

More precisely, according to [9,10], an analogical proportion (AP) on a set X is a quaternary relation on X, i.e., a subset of X^4. An element of this subset, written $x : y :: z : t$, is read as 'x is to y as z is to t', must obey the following two axioms:

1. Symmetry of "as": $x : y :: z : t \Leftrightarrow z : t :: x : y$
2. Exchange of means: $x : y :: z : t \Leftrightarrow x : z :: y : t$.

Analogical proportions [9,10] can be formulated for numbers, sets, in the boolean case, strings, as well as in various algebraic structures, like semigroups or lattices. In the latter case, we say that four elements (or in the case of a concept lattice, formal concepts) (x, y, z, t) of a lattice are in a *Weak Analogical Proportion (WAP)* iff $x \vee t = y \vee z$ and $x \wedge t = y \wedge z$. The symbols $\vee$, respectively $\wedge$, denote the supremum, respectively the infimum as they are defined in Order Theory. Therefore, if we denote by O_x and A_x the extent, respectively the intent of concept x, then, by the main theorem of FCA, the previous conditions are equivalent to

$$A_x \cap A_t = A_y \cap A_z \text{ and } O_x \cap O_t = O_y \cap O_z.$$

In some cases, when one of the four elements of the proportion is not known, it can be inferred from the three other elements. The idea of analogy is to establish a parallel between two situations, i.e., what is true in the first situation may also be true in the second one. However, the parallel between two situations that refer to apparently unrelated domains may be especially rich.

Let now (G, M, I) be a formal context, $O_1, O_2, O_3, O_4 \subseteq G$ sets of objects, $A_1, A_2, A_3, A_4 \subseteq M$ sets of attributes. We define the following subcontext, where X denotes that *all* elements of the respective subsets are related by the incidence relation I (Table 1):

Table 1. Formal context of concepts

	A_1	A_2	A_3	A_4
O_1			X	X
O_2		X		X
O_3	X		X	
O_4	X	X		

When this pattern is maximal, the subcontext is called *analogical complex*. Moreover, every analogical complex defines a WAP and viceversa. Given an analogical complex, let us denote by x, y, z and t the four concepts building a WAP. Analogical complexes contain correspondences between formal concepts as a whole, i.e., elements of the concept lattice. However, if we need to display analogies between objects and attributes, then another approach, called *proportional analogies* is required. For this, Miclet et al. introduce so called *full weak analogical proportions*, which are a more particular case of WAPs, eliminating what they think are trivial WAPs. A full weak analogical proportion (FWAP) [9] is a WAP (x, y, z, t) iff the four concepts are incomparable for $\leq$ and the quadruples (x, y, z, t) and (x, z, y, t) are not WAPs.

A *proportional analogy between concepts* [9] is a relation $\updownarrow$ defined on $\mathfrak{P}(G) \times \mathfrak{P}(M) \times \mathfrak{P}(G) \times \mathfrak{P}(M)$ written as

$$(O_x \setminus O_y) \updownarrow (A_x \setminus A_y) \updownarrow\updownarrow (O_y \setminus O_x) \updownarrow (A_y \setminus A_x).$$

and derived from a full analogical proportion between concepts. However, in this paper we use a "weaker" version of proportional analogies, by taking into account proportional analogies derived from WAPs and not just FWAPS.

A heuristic algorithm to discover such proportions by inspecting a lattice of formal concepts has been proposed in [10]. We have used analogical proportions and analogical complexes in order to compare objects with respect to their attribute values. Our research is following the analogy-based decision approach, through which medical decisions may be inferred from previously formed medical analogies.

3   Motivation

Data mining results are typically difficult to interpret, and much effort is necessary for domain experts to turn the results to practical use. In general, users do not care how sophisticated a data mining method is, but they do care how understandable its results are. Therefore, no method is acceptable in practice unless it's results are understandable [7].

The motivation of our research is to use analogical reasoning in order to improve the diagnosis process and to lower the risk of human error. Let's consider the following situation in order to have a better understanding of how analogical reasoning can be of use. A doctor of a certain specialization can only diagnose patient with regard to diseases from that particular specialization. In reality, it is often the case that a patient has a disease at the borderline of two or more specializations. What usually happens in that case is that the patient is sent from one doctor to another and he is subjected to multiple tests and investigations. In the end he might even end up with the wrong diagnosis because the two doctors rarely come together to discuss the case an each doctor has limited knowledge about other specializations. How can analogical reasoning help improve this situation?

Having a large database with patients, diseases and symptoms, we can compute proportional analogies, for instance of the form "symptom A is to disease B as symptom C is to disease D". Assuming that symptom A and disease B refer to the specialization of a doctor, it follows that he knows exactly what the correlation is between symptom A and disease B, for example it is the main symptom of that disease, or it is a secondary symptom of that disease or maybe they are not correlated at all. With this knowledge, the doctor can infer new information based on the analogy, namely how symptom C is related to disease D.

Another example would be the following proportional analogy obtained from the analyzed data: *Osteosclerosis* is to *Autophony, Hearing Loss, Rhinorrhea* what *Acute Pharyngitis* is to *Dysphagia, Fever*. A doctor specialized in throat diseases would know that *Dysphagia* and *Fever* are the principal symptoms of *Acute Pharyngitis*. Therefore, he can conclude that a patient having *Autophony, Hearing Loss* and *Rhinorrhea* as symptoms can be diagnosed with *Osteosclerosis* based on analogical reasoning.

In Sect. 4 we show how to compute weak analogical proportions and their corresponding proportional analogies for a given dataset. Furthermore, we present a few of the obtained results to show how analogical reasoning can be extended on the whole dataset.

4   Experiments

Medical diagnosis is considered one of the most important, and at the same time, complicated task, that needs to be executed accurately and efficiently. The development of an automatic system which may be used as a decision support system which can aid clinicians is therefore of high importance in order to improve accuracy and efficiency.

There are multiple situations in which valuable knowledge is hidden in some medical datasets, but it is not directly accessible and quite often this knowledge remains completely inaccessible to medical experts, which need to rely on their expertize, knowledge, experience and sometimes intuition. Of course, this does not necessarily mean that the diagnosis is wrong, but the purpose it may sometimes lead to unwanted biases, errors and excessive medical costs which affects the quality of service provided to patients [15]. Data mining techniques have the potential to generate a knowledge-rich environment which can help to significantly improve the quality of clinical decisions. The purpose of such tools is not to replace the doctors' experience and intuition, but to support their decision with facts and knowledge extracted from the data.

We are considering for this analysis records collected from a teaching hospital in Romania. The collected data refer to personal characteristics, symptoms and diagnostics of patients who came to the hospital for an investigation in the Department of Otorhinolaryngology. This department is specialized in the diagnosis and treatment of ear, nose and throat disorders.

Using these data, we show how new knowledge about medical investigations can be discovered, by following several steps: finding concepts, building knowledge concept lattices, finding relations and analogical proportions between concepts.

For a better understanding, we detect and represent the analogical proportions on a concept lattice, which has an intuitive and clear representation. Figure 1 presents the conceptual scale for the relation between diagnostics and symptoms which occurred for both male and female patients, but with a particular focus on *Deviated Septum* as principal diagnostic and *Chronic Sinusitis* as secondary diagnostic together with the corresponding symptoms that might appear in patients' data. We are not directly interested in the correlation between the gender of the patient and the diagnostics, but in the cause that led to the diagnostics. For that reason, we look for analogical structures in the concept lattice in order to infer knowledge regarding the correlation between symptoms and diagnostics.

Following the notions introduced in the preliminaries, we look for formal concepts in the lattice that form WAPs, i.e. quadruples of concepts (x, y, z, t), where the two pairs (x, t) and (y, z) share the same infimum and supremum. In this particular case, we identified four quadruples of concepts that fulfill this property. The four WAPs can be observed in Figs. 1a–c, respectively Fig. 1d, where each WAP is of the form (x, y, z, t), with x, y, z and t being formal concepts (and therefore represented as nodes in the concept lattice). The pairs of concepts sharing the same infimum and supremum are highlighted with different borders for a clear view of the weak analogical proportion.

As explained in the preliminaries, in order to obtain correspondences between diagnostics (objects in the represented formal context) and symptoms (attributes in the represented formal context) we need to compute the corresponding proportional analogies. Herefrom, we can infer similarities between the relations linking different subsets objects to different subsets of attributes. In order to do this we need a clear view of the extents and intents of each concept. Therefore, although the extents and intents can be read directly from the concept lattice, in order to make it easier to follow the computations in the next step, we display them separately in tables. These can be found in Tables 2, 3, 4 and 5 for WAP1 - WAP4, respectively. We observe that WAP2 is only a permutation of WAP1 (obtained by exchanging concepts t and z), hence the formal concepts are the same as in Table 2, but with different notations.

When computing the proportional analogies, the definition given by Miclet et al. states that from the WAP formed by the concepts x, y, z, and t one can compute the proportional analogy for the pair (x, y). However, given the commutativity of the supremum and infimum operations, we can obviously deduce the same proportional analogies for the following pairs: $(x, z), (y, t)$, and (z, t). As described in the preliminaries, we compute the proportional analogies as set differences of the extents and intents of the detected pairs.

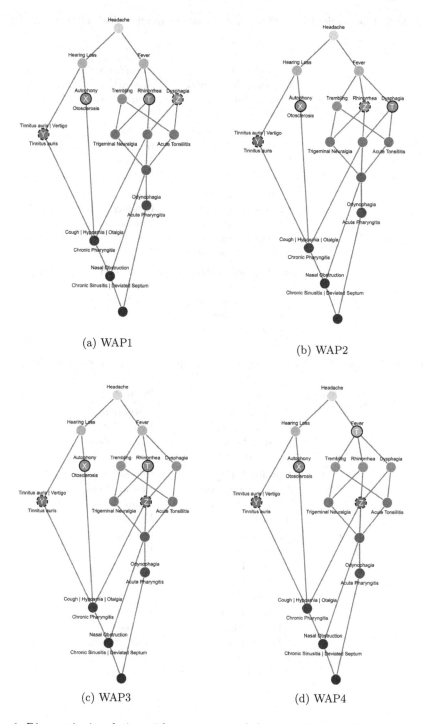

(a) WAP1

(b) WAP2

(c) WAP3

(d) WAP4

Fig. 1. Diagnostics in relation with symptoms with focus on Deviated Septum as principal diagnostic and Chronic Sinusitis as secondary diagnostic

Table 2. Diagnostic: Deviated Septum - concepts in analogical relation - WAP1

Concept	Extent	Intent
x	Ostosclerosis Chronic Pharyngitis Chronic Sinusitis Deviated Septum	Autophony Headache Hearing Loss
y	Chronic Pharyngitis Chronic Sinusitis Deviated Septum Tinnitus	Headache Hearing Loss Tinnitus Vertigo
t	Acute Pharyngitis Chronic Pharyngitis Chronic Sinusitis Deviated Septum Trigeminal Neuralgia	Fever Headache Rhinorrhea
z	Acute Pharyngitis Acute Tonsilitis Chronic Pharyngitis Chronic Sinusitis Deviated Septum	Dysphagia Fever Headache

Table 3. Diagnostic: Deviated Septum - concepts in analogical relation - WAP2

Concept	Extent	Intent
x	Ostosclerosis Chronic Pharyngitis Chronic Sinusitis Deviated Septum	Autophony Headache Hearing Loss
y	Chronic Pharyngitis Chronic Sinusitis Deviated Septum Tinnitus	Headache Hearing Loss Tinnitus Vertigo
t	Acute Pharyngitis Acute Tonsilitis Chronic Pharyngitis Chronic Sinusitis Deviated Septum	Dysphagia Fever Headache
z	Acute Pharyngitis Chronic Pharyngitis Chronic Sinusitis Deviated Septum Trigeminal Neuralgia	Fever Headache Rhinorrhea

Table 4. Diagnostic: Deviated Septum - concepts in analogical relation - WAP3

Concept	Extent	Intent
x	Ostosclerosis Chronic Pharyngitis Chronic Sinusitis Deviated Septum	Autophony Headache Hearing Loss
y	Chronic Pharyngitis Chronic Sinusitis Deviated Septum Tinnitus	Headache Hearing Loss Tinnitus Vertigo
t	Acute Pharyngitis Chronic Pharyngitis Chronic Sinusitis Deviated Septum Trigeminal Neuralgia	Fever Headache Rhinorrhea
z	Acute Pharyngitis Acute Tonsilitis Chronic Pharingitis Chronic Sinusitis Deviated Septum	Dysphagia Fever Headache Rhinorrhea

Table 5. Diagnostic: Deviated Septum - concepts in analogical relation - WAP4

Concept	Extent	Intent
x	Ostosclerosis Chronic Pharyngitis Chronic Sinusitis Deviated Septum	Autophony Headache Hearing Loss
y	Chronic Pharyngitis Chronic Sinusitis Deviated Septum Tinnitus	Headache Hearing Loss Tinnitus Vertigo
t	Acute Pharyngitis Chronic Pharyngitis Chronic Sinusitis Deviated Septum	Dysphagia Fever Headache Rhinorrhea
z	Acute Pharyngitis Acute Tonsilitis Chronic Pharingitis Chronic Sinusitis Deviated Septum Trigeminal Neuralgia	Fever Headache

Table 6. Diagnostic: Deviated Septum - proportional analogies - WAP1

	Extent1\Extent2	Intent1\Intent2	Extent2\Extent1	Intent2\Intent1
(x, y)	Ostosclerosis	Autophony	Tinnitus	Tinnitus Vertigo
(x, z)	Ostosclerosis	Autophony Hearing Loss	Acute Pharyngitis Acute Tonsilitis	Dysphagia Fever
(y, t)	Tinnitus	Hearing Loss Tinnitus Vertigo	Acute Pharyngitis Trigeminal Neuralgia	Fever Rhinorrhea

Table 7. Diagnostic: Deviated Septum - proportional analogies - WAP2

	Extent1\Extent2	Intent1\Intent2	Extent2\Extent1	Intent2\Intent1
(x, z)	Ostosclerosis	Autophony Hearing Loss	Acute Pharyngitis Trigeminal Neuralgia	Fever Rhinorrhea
(y, t)	Tinnitus	Hearing Loss Tinnitus Vertigo	Acute Pharyngitis Acute Tonsilitis	Dysphagia Fever

Table 8. Diagnostic: Deviated Septum - proportional analogies - WAP3

	Extent1\Extent2	Intent1\Intent2	Extent2\Extent1	Intent2\Intent1
(x, z)	Ostosclerosis	Autophony Hearing Loss Rhinorrhea	Acute Pharyngitis	Dysphagia Fever
(y, t)	Tinnitus	Hearing Loss Tinnitus Vertigo	Acute Pharyngitis Trigeminal Neuralgia	Fever Rhinorrhea

Table 9. Diagnostic: Deviated Septum - proportional analogies - WAP4

	Extent1\Extent2	Intent1\Intent2	Extent2\Extent1	Intent2\Intent1
(x, z)	Ostosclerosis	Autophony Hearing Loss Rhinorrhea	Acute Pharyngitis	Dysphagia Fever
(y, t)	Tinnitus	Hearing Loss Tinnitus Vertigo	Acute Pharyngitis Trigeminal Neuralgia Acute Tonsilitis	Fever

Given these considerations, the proportional analogies for WAP1–WAP4 are displayed in Tables 6, 7, 8 and 9, respectively. Trivial WAPs, where one of the components is the empty set were eliminated from this analysis.

In this small and relatively easy to read example, we present a few of the weak analogical proportions and their corresponding proportional analogies in order to prove how specific domain knowledge that can be inferred from analogical reasoning. For instance, by analyzing the pair (x, z) from Table 6 and the pair (x, z) from Table 7, we can observe that the pair of objects and attributes {{*Ostoclerosis*}, {*Autophony, Hearing Loss*}} is repeated, even if considering different weak analogical proportions. However the second part of the proportional analogy is different, so we can infer different knowledge and enhance the chain of similarities. After analyzing the pair (y, t) from Table 6 and the pair (y, t) from example Table 7 we can observe that if *Rhinorrhea* is replaced with *Dysphagia* in the symptoms list, there will be replacements in the diagnostics list (*Acute Tonsilitis* will be replaced by *Trigeminal Neuralgia*).

By this small example, we can remark from the data under analysis, on the one hand the real value of the analogical reasoning in this medical setting. For instance, one can deduce that a group of symptoms might describe in a unique way a set of diagnostics. On the other, changing the choice of a single symptom can eventually influence the diagnostic of the patient. For a medical expert, these facts can lead to a better understanding of various patient related data or even to discover eventual inconsistencies in the data under analysis. Moreover, pairs of concepts in weak analogical proportions can be used, combined with the corresponding concept lattice where they are highlighted, for further communication, as an inference support or even as a knowledge discovery method. Facts that stand out should be analyzed on patients' records over a large period of time and, if they persist, it can lead to the formulation of some hypotheses which can than be researched in more details by medical staff for a validation.

Electronic Health Records with an analogical reasoning component enable the discovery of in-depth knowledge which is not always easy to discover. Such a system may address recommendation tasks in order to support clinicians in their medical decisions. Through a small scenario we want to emphasize how a clinician, who has medical experience, can interact with an Electronic Health Record with an analogical reasoning component grounded on FCA:

- Patients comes to the hospital and he/she presents a list of symptoms which are recorded in the system;
- Some of the patient's symptoms are principal symptoms, while others are secondary symptoms;
- Based on a previous data preprocessing phase, conceptual landscapes are built, enabling navigation in a dyadic or triadic setting;
- The analogical reasoning component presents quadruples of concepts in a weak analogical proportion and mines all analogical relations;
- Based on a previously built knowledge base, eventual inconsistencies can be highlighted;
- The clinician can give a diagnostic to the patient and investigates some of the highlighted analogical relations;
- Even more, based on the entire set of analogies, the clinician may dig deeper into relations between symptoms and diagnostics and he/she can specify

exactly which diagnostic is the main diagnostic and which are the secondary diagnostics based on the specific list of symptoms.

5   Conclusions and Future Work

Electronic Health Record systems proved to be a significantly aid in health improvement and also a continuous challenge. By means of FCA and analogical reasoning we succeeded in addressing health care system's shortcomings and we have proved that they might have a positive impact on patient welfare.

Through analogical reasoning medical errors can be decreased and due to the existing analogies between patient symptoms and diagnostics, preventing care of a patient may be enhanced.

As future work we propose to develop a recommendation system based on the confidence of an analogy which may be a real help for clinicians. Our perspective is moving on assigning a grading to the recommendation according to the percent of concepts in which the pair (*object, attribute*) is present out of all concepts which contain the *object* in their extent.

References

1. Correa, W.F., Prade, H., Richard, G.: Trying to understand how analogical classifiers work. In: Hüllermeier, E., Link, S., Fober, T., Seeger, B. (eds.) SUM 2012. LNCS (LNAI), vol. 7520, pp. 582–589. Springer, Heidelberg (2012). https://doi.org/10.1007/978-3-642-33362-0_46
2. Delen, D., Walker, G., Kadam, A.: Predicting breast cancer survivability: a comparison of three data mining methods. Artif. Intell. Med. **34**(2), 113–127 (2005)
3. Ganter, B., Wille, R.: Formal Concept Analysis - Mathematical Foundations. Springer, Heidelberg (1999). https://doi.org/10.1007/978-3-642-59830-2
4. Haliță, D., Săcărea, C.: Is FCA suitable to improve electronic health record systems? In: SoftCom 2016, pp. 1–5. IEEE (2016)
5. Hillestad, R., Bigelow, J., Bower, A., Girosi, F., Meili, R., Scoville, R., Taylor, R.: Can electronic medical record systems transform health care? Potential health benefits, savings, and costs. Health Aff. **24**(5), 1103–1117 (2005)
6. Jay, N., Kohler, F., Napoli, A.: Using formal concept analysis for mining and interpreting patient flows within a healthcare network. In: Yahia, S.B., Nguifo, E.M., Belohlavek, R. (eds.) CLA 2006. LNCS (LNAI), vol. 4923, pp. 263–268. Springer, Heidelberg (2008). https://doi.org/10.1007/978-3-540-78921-5_19
7. Li, J., Fu, A., He, H., Chen, J., Jin, H., McAullay, D., Williams, G., Sparks, R., Kelman, C.: Mining risk patterns in medical data. In: SIGKDD 2005, pp. 770–775. ACM (2005)
8. Menachemi, N., Collum, T.: Benefits and drawbacks of electronic health record systems. Risk Manag. Healthc. Policy **5**, 47–55 (2011)
9. Miclet, L., Barbot, N., Prade, H.: From analogical proportions in lattices to proportional analogies in formal concepts. In: Schaub, T., Friedrich, G., O'Sullivan, B. (eds.) ECAI 2014–21st European Conference on Artificial Intelligence, 18–22 August 2014, Prague, Czech Republic - Including Prestigious Applications of Intelligent Systems (PAIS 2014), Frontiers in Artificial Intelligence and Applications, vol. 263, pp. 627–632. IOS Press (2014)

10. Miclet, L., Nicolas, J.: From formal concepts to analogical complexes. In: Proceedings of Twelfth International Conference on Concept Lattices and Their Applications, Clermont-Ferrand, France, 13–16 October 2015, pp. 159–170 (2015)
11. Miclet, L., Prade, H., Guennec, D.: Looking for analogical proportions in a formal concept analysis setting. In: Proceedings of 8th International Conference on Concept Lattices and Their Applications, Nancy, France, 17–20 October 2011, pp. 295–307 (2011)
12. Săcărea, C.: Investigating oncological databases using conceptual landscapes. In: Hernandez, N., Jäschke, R., Croitoru, M. (eds.) ICCS 2014. LNCS (LNAI), vol. 8577, pp. 299–304. Springer, Cham (2014). https://doi.org/10.1007/978-3-319-08389-6_26
13. Săcărea, C., Sotropa, D., Troanca, D.: Symptoms investigation by means of formal concept analysis for enhancing medical diagnoses. In: Begusic, D., Rozic, N., Radic, J., Saric, M. (eds.) 25th International Conference on Software, Telecommunications and Computer Networks, SoftCOM 2017, Split, Croatia, 21–23 September 2017, pp. 1–5 IEEE (2017)
14. Slight, S.P.: Meaningful use of electronic health records: experiences from the field and future opportunities. JMIR Med. Inform. 3(3), 1–11 (2015)
15. Soni, J., Ansari, U., Sharma, D., Soni, S.: Predictive data mining for medical diagnosis: an overview of heart disease prediction. Int. J. Comput. Appl. 17(8), 43–48 (2011)
16. Sowa, J., Majumdar, A.: Analogical reasoning. In: Conceptual Structures for Knowledge Creation and Communication, 11th International Conference on Conceptual Structures, ICCS 2003 Dresden, Germany, 21–25 July 2003 Proceedings, pp. 16–36 (2003)
17. Tang, P.C., McDonald, C.J.: Electronic Health Record Systems. In: Shortliffe, E.H., Cimino, J.J. (eds.) Biomedical Informatics: Computer Applications in Health Care and Biomedicine. HI. Springer, New York (2006). https://doi.org/10.1007/0-387-36278-9_12
18. Wille, R.: Conceptual contents as information - basics for contextual judgment logic. In: Conceptual Structures for Knowledge Creation and Communication, Proceedings of 11th International Conference on Conceptual Structures, ICCS 2003 Dresden, Germany, 21–25 July 2005, pp. 1–15 (2003)
19. Wille, R.: Methods of conceptual knowledge processing. In: Missaoui, R., Schmidt, J. (eds.) ICFCA 2006. LNCS (LNAI), vol. 3874, pp. 1–29. Springer, Heidelberg (2006). https://doi.org/10.1007/11671404_1

An Efficient Approximation of Concept Stability Using Low-Discrepancy Sampling

Mohamed-Hamza Ibrahim$^{(\boxtimes)}$ and Rokia Missaoui

Département d'informatique et d'ingénierie, Université du Québec en Outaouais,
101, rue St-Jean Bosco, Gatineau, Québec J8X 3X7, Canada
{ibrm05,rokia.missaoui}@uqo.ca

Abstract. One key challenge in Formal Concept Analysis is the scalable and accurate computation of stability index as a means to identify relevant formal concepts. Unfortunately, most exact methods for computing stability have an algorithmic complexity that could be exponential w.r.t. the context size. While randomized approximate algorithms, such as Monte Carlo Sampling (MCS), can be good solutions in some situations, they frequently lead to the slow convergence problem with an inaccurate estimation of stability. In this paper, we introduce a new approximation method to estimate the stability using the low-discrepancy sampling (LDS) approach. To improve the convergence rate, LDS uses quasi-random sequence to distribute the sample points evenly across the power set of the concept intent (or extent). This helps avoid the clumping of samples and let all the areas of the sample space be duly represented. Our experiments on several formal contexts show that LDS can achieve faster convergence rate and better accuracy than MCS.

Keywords: Formal Concept Analysis · Concept stability
Low-discrepancy sampling · Pattern relevancy

1 Introduction

Formal Concept Analysis (FCA) [10] is a theoretical framework based on lattice and order theory [10], which is a formalization of concept lattice and hierarchy. The concept lattice contains frequently substantial combinatorial structures that are exploited for data mining and analysis purposes. However, a large amount of such structures could be irrelevant, which in turn, potentially induce a high complexity even for small datasets [4,11]. Thus, an important challenge in the big data era is to discover only relevant concepts from a possibly very large and complex lattice. Inspired by the FCA theory,, concept selection techniques are commonly used to focus on only relevant parts of a concept lattice [16,17]. The basic idea is to single out a subset of important concepts, objects or attributes based on relevance measures. Several selection measures have been introduced as

© Springer International Publishing AG, part of Springer Nature 2018
P. Chapman et al. (Eds.): ICCS 2018, LNAI 10872, pp. 24–38, 2018.
https://doi.org/10.1007/978-3-319-91379-7_3

clearly detailed in [17], such as stability [13,15], separation [11], probability [16] and robustness [11]. While each selection measure exploits certain valuable characteristics in the formal context, stability has been found to be more prominent for assessing concept relevancy [16].

The stability index of the concept $c = (A, B)$ quantifies how its intent B depends on the set of objects in its extent A. Thus, the stability frequently provides a robust indication of how much noise appears in the concept. Unfortunately, computing the stability is #P-complete [2,15] approximating, and we typically need to compute the set of all generators associated with the concept to get its exact stability value. This could be problematic with large-sized concepts, and thus, a practical solution to this problem could be the stability approximation. The authors in [2] introduced random MCS as a way to approximate the stability. Although MCS might give a good approximation in some given cases [18], it converges very slowly and needs more samples to reduce the sampling error. This is due to the fact that MCS randomly chooses samples independently. This leads MCS to end up with some regions in the power set space of the concept intent (or extent) with too many samples tightly clumped, while other regions have no samples. This weakens the convergence rate[1] of the sampling process and worsen the inaccuracy of stability [21,22]. In this paper, our main objective is to use a low-discrepancy approach [7,24] to address the limitation of MCS.

In the following we first give a background, then we explain our general LDS approach in Sect. 3 and illustrate its application to stability approximation. Finally, we present experimental evaluations in Sect. 4, followed by our conclusion in Sect. 5.

2 Background

2.1 Formal Concept Analysis

FCA is a mathematical formalism for data analysis [10] that uses a formal context as input to construct a set of formal concepts organized in a concept lattice. A *formal context* is a triple $\mathbb{K} = (\mathcal{G}, \mathcal{M}, \mathcal{I})$, where $\mathcal{G}$ is a set of objects, $\mathcal{M}$ a set of attributes, and $\mathcal{I}$ a relation between $\mathcal{G}$ and $\mathcal{M}$ with $\mathcal{I} \subseteq \mathcal{G} \times \mathcal{M}$. For $g \in \mathcal{G}$ and $m \in \mathcal{M}, (g, m) \in \mathcal{I}$ holds iff the object g has the attribute m. Given arbitrary subsets $A \subseteq \mathcal{G}$ and $B \subseteq \mathcal{M}$, the Galois connection can be defined by the following derivation operators:

$$A' = \{m \in \mathcal{M} \mid \forall g \in A, (g, m) \in \mathcal{I}\}, \ A \subseteq \mathcal{G}$$

$$B' = \{g \in \mathcal{G} \mid \forall m \in B, (g, m) \in \mathcal{I}\}, \ B \subseteq \mathcal{M}$$

where A' is the set of attributes common to all objects of A and B' is the set of objects sharing all attributes from B. The closure operator $(.)''$ implies the double application of $(.)'$, which is extensive, idempotent and monotone. The

---

[1] The convergence rate quantifies how quickly the sampling error decreases with an increase in the number of samples.

subsets A and B are said to be closed when $A'' = A$, and $B'' = B$. If both A and B are closed and $A' = B$, and $B' = A$, then the pair $c = (A, B)$ is called a *formal concept* of $\mathbb{K}$ with *extent* A and *intent* B. For a finite intent (or extent) set of w elements, we use $\mathcal{P}(.)$ to denote its power set with a number of subsets equals to $n = 2^w$, i.e., the set of all its subsets, including the empty set and the set itself.

2.2   Stability Index

The stability index $\sigma(c)$ is an interestingness measure of a formal concept c for selecting relevant patterns [5,11,15,23].

Definition 1. *Let* $\mathbb{K} = (\mathcal{G}, \mathcal{M}, \mathcal{I})$ *be a formal context and* $c = (A, B)$ *a formal concept of* $\mathbb{K}$. *The intensional stability* $\sigma_{in}(c)$ *can be computed as [2,5]:*

$$\sigma_{in}(c) = \frac{|\{e \in \mathcal{P}(A)|e' = B\}|}{2^{|A|}} \tag{1}$$

In a dual way, the extensional stability $\sigma_{ex}(c)$ *is defined as:*

$$\sigma_{ex}(c) = \frac{|\{e \in \mathcal{P}(B)|e' = A\}|}{2^{|B|}} \tag{2}$$

In Eq. (1), intensional stability $\sigma_{in}(c)$ measures the strength of dependency between the intent B and the objects of the extent A. More precisely, it expresses the probability to maintain B closed when a subset of noisy objects in A are deleted with equal probability. The numerator of Eq. (1) can be calculated by naively iterating through all the subsets in $\mathcal{P}(A)$ and counting the number of those subsets whose derivation operator produces the intent B. This computation needs a time complexity of $O(2^{|A|})$. However, it had been shown in [23] and later on in [25] that the numerator of $\sigma_{in}(c)$ in Eq. (1) can be computed by identifying and counting the minimal generators of the concept. Such computation takes a time complexity of $O(L^2)$ [23,25], where L is the size of the concept lattice, and requires the lattice construction which needs a time complexity of $O(|\mathcal{G}|^2 \cdot |\mathcal{M}| \cdot L)$ [14,20]. In a dual way, the extensional stability $\sigma_{ex}(c)$ (see Eq. (2)) measures overfitting in the intent B. When it is high for the extent A of a given concept c, it means that A would stay closed even when we discard noisy attributes from its intent B. Similar to the intensional stability, $\sigma_{ex}(c)$ can be computed by simply iterating through all the subsets in $\mathcal{P}(B)$ but counting the number of those subsets for which the derivation operator produces the extent A. This computation also needs exponential time complexity $O(2^{|B|})$.[2]

---

[2] Note that since both $\sigma_{in}(c)$ and $\sigma_{ex}(c)$ provide dual measurements, we generally use $\sigma(c)$ throughout the rest of the paper.

2.3   Approximating Stability

With a large-sized intent (or extent), the exponential time complexity of computing the stability represents a bottleneck for any exact method. For instance, to exactly compute $\sigma(c)$ of a concept with an intent size $n = 21$, we need to perform more than two million computational comparisons. Thus, a practical solution is to approximate the stability using MCS method [2,6] as shown in Algorithm 1. Under a certain number of samples N, MCS iteratively picks up a random subset from the intent power set $\mathcal{P}(B)$ and increments a counter when the picked up subset satisfies the condition in the numerator of Eq. (2) (see lines 3–5). It ultimately uses the counter to estimate the stability (see line 8). However, how many samples are needed in the MCS to accurately estimate the stability? In [2,5] the authors demonstrated that at least a sample size of $N > \frac{1}{2\epsilon^2} \ln \frac{2}{\delta}$ could be sufficient for MCS to estimate the stability with precision ϵ and error rate δ. This means that, with a probability $1 - \delta$, we can calculate an approximation $\tilde{\sigma}(c)$ of the exact stability $\sigma(c)$ within the interval $[\tilde{\sigma}(c) - \epsilon, \tilde{\sigma}(c) + \epsilon]$. At a first sight, this appears to be a low-cost computational time compared to the exact computational one, but in fact it is not. For example, MCS needs a sample size $N > 3.8 \times 10^6$ to accurately estimate the stability with precision $\epsilon = 0.001$ and error rate $\delta = 0.001$.

Algorithm 1. MCS for stability approximation

Input: Concept $c = (A, B)$ and Sample size N.
Output: Estimated stability $\tilde{\sigma}(c)$.
 1: Count $\leftarrow$ 0;
 2: **for** $i \leftarrow 1$ to N **do**
 3:     Pick up a random subset e in $\mathcal{P}(B)$;
 4:     **if** $e' == A$ **then**
 5:         Count $\leftarrow$ Count + 1;
 6:     **end if**
 7: **end for**
 8: $\tilde{\sigma}(c) \leftarrow \frac{\text{Count}}{N}$;

---

3   LDS for Stability

In this section we will explain the usage of the LDS method for estimating stability.

3.1   LDS Framework

Let us first reformulate the problem of computing stability so that the LDS sampling can be easily comprehended. Given a formal concept c with extent A and intent B, then the power set of B, $\mathcal{P}(B) = \{b_{x_1}, \ldots, b_{x_n}\}$ is associated with a set of indices $X = \{x_1, \ldots, x_n\}$, where $x_i \in X$ is the index of $b_{x_i} \in \mathcal{P}(B)$. We define $\mathbb{1}(x)$ as an indicator function over X as follows:

$$\mathbb{1}(x) = \begin{cases} 1 & \text{If } b'_x = A, \ b_x \in \mathcal{P}(B) \\ 0 & \text{Otherwise} \end{cases} \tag{3}$$

where $\mathbb{1}(x)$ is equal to 1 when a randomly picked up subset $b_x \in \mathcal{P}(B)$ satisfies the condition in the numerator of stability in Eq. (2); and 0 otherwise. Note that the concept stability in Eq. (2) is identical to the following quantity:

$$\sigma(c) \equiv p(b'_x = A \mid b_x \in \mathcal{P}(B)) \tag{4}$$

where $p(b'_x = A \mid b_x \in \mathcal{P}(B))$ is the probability of randomly picking up a subset b_x and finding out that it satisfies the stability condition. This probability can be computed by using the expectation $E[.]$ of the indicator function $\mathbb{1}(x)$ in Eq. (3) as follows:

$$p(b'_x = A \mid b_x \in \mathcal{P}(B)) = E[\mathbb{1}(x)] = \sum_{x \in X} \mathbb{1}(x) P_X(x) \tag{5}$$

where $P_X(x)$ is the (univariate) probability mass function of X. If we assume that the subsets in $\mathcal{P}(B)$ are uniformly distributed, then

$$P_X(x) = \frac{1}{n}, \ \forall x \in X \tag{6}$$

where $n = |X|$ is the size of the power set. Now if we use Eqs. (4) and (6) into (5), then the stability in Eq. (2) can be computed as:

$$\sigma(c) = E[\mathbb{1}(x)] = \frac{1}{n} \sum_{x \in X} \mathbb{1}(x) \tag{7}$$

The stability in Eq. (7) can be approximated by taking a set of samples $\mathcal{S}$ as:

$$\sigma(c) \approx \tilde{\sigma}(c; N) = \frac{1}{N} \sum_{x_s \in \mathcal{S}} \mathbb{1}(x_s) \tag{8}$$

where $N = |\mathcal{S}|$ is the sample size. Now, how can we select correlated N samples (i.e., subsets), in Eq. (8), that are distributed across the power set space more uniformly than uncorrelated random samples in MCS? The answer is the usage of low-discrepancy sequence. In the following subsection, let us explain how to generate low-discrepancy (also called quasi-random) sample points.

3.2   Generating Low-Discrepancy Sequences

Various deterministic methods can be typically used to generate low-discrepancy sequences of N sample points $\{s_i\}_{i=1}^N$. Here we focus on the two commonly used sequences called *Scrambled Van der Corput* [9,24] and *Sobol* [12,19].

(I) Scrambled Van der Corput Sequence (VDC) is the simplest low discrepancy sequence. Starting with a prime integer $r \geq 2$ as a radical base to represent the sequence, we can obtain the ith sample point s_i by calling the function `GenerateSVdC(i,r)` in Algorithm 2. We first represent the decimal number i in radical base r (line 1). Then, we put a radical point in the front of its reversal representation and convert it back to the decimal (Line 2). We finally scramble the result by shifting it with a pseudo-random number (see lines 2–4). For example, if we choose base $r = 2$ in Scrambled VDC, then we can get the 4th sample point s_4 as follows: (1) express $i = 4$ in base 2 as $(100)_2$; (2) reverse the digits and put a radix point in front of it to get $(0.001)_2$; (3) convert it back to a decimal number to get $h_4 = 0.375$; (4) scramble h_4 with any random number (say $d_4 = 0.8$) to obtain $s_4 = 0.375 + 0.8 \ (\text{mod } 1) = 0.175$.

Algorithm 2. Generating a sample point in Scrambled VDC.

Function `GenerateSVdC`
Input: Number i and radical base r
Output: A point $s_i \in [0, 1]$

    // Express i as a number in base r.
1: $i \rightarrow \sum_{j=0}^{l} a_j r^j$;
    // Reverse the digits, put a radix point and convert it back to decimal.
2: $h_i \leftarrow \sum_{j=0}^{l} a_j r^{-j-1}$; // $l = \lfloor \log_r i \rfloor$.
    // Scramble the number by randomized shifting.
3: $d_i \leftarrow$ Generate a uniform random number $\in [0, 1]$;
4: $s_i \leftarrow h_i + d_i \ (\text{mod } 1)$;
5: **Return** s_i;

(II) One-dimensional Sobol Sequence is often generated based on radical base $r = 2$. Unlike Scrambled VDC with base 2, it applies the permutations based on a set of direction numbers (instead of reversing numbers). The function `GenerateSobol(i)` in Algorithm 3 summarizes the pseudo-code of Sobol for generating a sample point s_i. We first compute the gray-code h_i of the number i by using the bit-by-bit exclusive-or operator (see line 1). Then we transform h_i to its binary representation (line 2). Subsequently, we sum up bit by bit exclusive-or of the direction numbers associated with the digits of h_i that are different from zero (see line 3). Note that the set of direction numbers ($\{v_j = \frac{z_j}{2^j}\}$, where $0 < z_j < 2^j$) is a sequence of binary fractions with bits after the binary point. Thus, they can be uniquely defined as $v_1 = (0.1)_2$, $v_2 = (0.11)_2$, $v_3 = (0.111)_2$, and so on. Now, as a concrete example, to obtain the 2nd sample point s_2 in the Sobol sequence, we do the following: (1) compute the gray code of $i = 2$ as $h_2 = 2 \oplus 1 = (10)_2 \oplus (1)_2 = (11)_2$; (2) apply exclusive-or to the first $v_1 = 0.1$ and the second $v_2 = 0.11$ direction numbers to get $u_2 = 0.1 \oplus 0.11 = 0.01$; (3) convert u_2 back to decimal to obtain Sobol point $s_2 = (0.01)_2 = 0.25$.

The limitation of q–dimensional Sobol sequence is that its convergence rate $O(\frac{(\log N)^q}{N})$ is smaller than $O(\frac{1}{\sqrt{N}})$ of the MCS convergence rate only when the problem dimension q is very small [19]. However, this limitation does not occur in our algorithms because the power set $\mathcal{P}(B)$ of an intent B has often one

Algorithm 3. Generating a sample point in Sobol sequence.

Function GenerateSobol
Input: Number i
Output: A point $s_i \in [0, 1]$

    // Compute the Gray Code of number i.
1: $h_i \leftarrow i \oplus \left\lceil \frac{i}{2} \right\rceil$;
    // Express h_i as a number in base 2.
2: $h_i \rightarrow \sum_{j=0}^{l} a_j 2^j$; // $l = \lfloor \log_r i \rfloor$.
    // Sum up the (XOR) of direction numbers associated with the digits of h_i.
3: $u_i \leftarrow a_1 v_1 \oplus \ldots \oplus a_l v_l$;
    // Put a radix point in the front of u_i and convert it back to decimal.
4: $s_i \leftarrow \sum_{j=0}^{l} a_j 2^{-j-1} \leftarrow u_i$;
5: **Return** s_i;

---

dimension (i.e., $q = 1$). Now, let us continue back to the approximation of stability in the next subsection.

3.3 LDS Algorithm for Estimating Stability

Here our goal is to improve the convergence rate by using LDS approach.

Algorithm 4 is the pseudo-code of the LDS procedure for approximating the stability depicted in Eq. (8). The algorithm takes as input the number of samples and the radical base, and then it applies two steps. In a first stage (see lines 1–6), LDS generates a low-discrepancy sequence S by using either the Scrambled VDC method (i.e., calling `GenerateSVdC(i,r)` in Algorithm 2) or Sobol method (i.e., calling `GenerateSobol(i)` in Algorithm 3). In a second stage, LDS exploits the obtained quasi-random sequence S to uniformly select correlated subsets across all areas of the intent power-set. That is, LDS iteratively uses each sample point $s_i \in S$ to pick up a corresponding subset index x_i by computing the inverse cumulative function $\mathcal{F}^{-1}(s_i)$ (see lines 9–10). Here, it is worth noting that because X is uniformly distributed (refer to Eq. (6)), then it has a discrete uniform cumulative distribution function $\mathcal{F}(X)$, in which its inverse $\mathcal{F}^{-1}()$ can be calculated as follows:

$$x_i = \mathcal{F}^{-1}(s_i) = \lceil s_i \times n \rceil \tag{9}$$

Then, it uses the indicator function (see Eq. (6)) to check the stability condition of the selected subset, and it consequently increments the *count* whenever the condition is held (see lines 11–14). Thus, *count* often captures the number of these selected subsets that satisfy the condition in the numerator of Eq. (8). Finally, the stability can be estimated as the portion of those subsets, from the N picked up ones, that satisfy the stability condition (line 15).

Algorithm 4. LDS for stability approximation.

Input: Concept $c = (A, B)$, sample size N and base r
Output: Stability $\tilde{\sigma}(c)$.

    `// Stage 1: Generate low-discrepancy sequence of points`
1: $S \leftarrow \emptyset$;
2: **for** $i \leftarrow 1$ to N **do**
3:    $s_i \leftarrow$ `GenerateSVdC(i,r);`   `// Case 1: using Scrambled VDC in Algo. 2.`
    `// OR`
4:    $s_i \leftarrow$ `GenerateSobol(i);`   `// Case 2: using sobol in Algo. 3.`
5:    $S \leftarrow S \cup \{s_i\}$;
6: **end for**
    `// Stage 2: Approximate the stability`
7: Count $\leftarrow 0$;
8: **for each** s_i in S **do**
    `// pick up a subset using the inverse cdf` $\mathcal{F}^{-1}(.)$
9:    $x_i \leftarrow \lceil s_i \times n \rceil$;
10:   $b_{x_i} \leftarrow$ Subset in $\mathcal{P}(B)$ at index x_i;
    `// Use` $\mathbb{1}(.)$ `in Eq. (3) to check if the subset satisfies the stability`
    `condition`
11:   **if** $\mathbb{1}(b_{x_i}) == 1$ **then**
12:     Count $\leftarrow$ Count $+ 1$;
13:   **end if**
14: **end for**
15: **Return** $\tilde{\sigma}(c) \leftarrow \frac{\text{Count}}{N}$;

3.4   LDS Versus MCS

In its basic form MCS method uses pseudo-random sequences. The main limitation that impacts its accuracy and scalability when estimating stability is the clumping that occurs when selecting the subsets based on the random or pseudo-random sequence. The reason for this clumping is that the different selected subsets know nothing about each other, which in turn makes them lie very close together in certain regions of the power set space while other regions have no selected subsets. As it has been shown in [22], about $\sqrt{N}$ out of N sample points could lie in clumps. This means that we need four times as many samples to halve the error, which dramatically influences the convergence rate. In fact, it has been proven that MCS method often provides a convergence rate of $O(\frac{1}{\sqrt{N}})$ using N samples [21]. Our proposed LDS method avoids the clumping by using low-discrepancy sequences instead of random or pseudo-random sequences. The former sequences often produce correlated samples in a quasi-random and deterministic fashion. Due to the correlations, those samples capture more uniformity properties and are less versatile than those samples of random or pseudo-random sequences. It has been proven in [7,21,22] that the usage of low-discrepancy sequences in sampling provides a resulting convergence rate of $O(\frac{\log N}{N})$. From a theoretical perspective, we believe that this significant convergence rate makes LDS often more accurate than MCS. For instance, after only $N = 100$ samples,

MCS could have a sampling error $O(\frac{1}{\sqrt{100}}) \approx 0.1$ whereas LDS has a sampling error $O(\frac{\log 100}{100}) \approx 0.02$.

4   Experimental Evaluation

The main goal of our experimental evaluation is to investigate the following key questions: (Q1) Is LDS more accurate than MCS for estimating stability even with small formal contexts? (Q2) Is LDS scalable when approximating the stability on large-sized concepts?

4.1   Methodology

We started our experiments by first selecting three real-life datasets: (1) *Mushroom* which describes mushroom species according to a set of features [3]; (2) *Phytotherapy* which presents diseases and medicinal plants that treat them [1]; (3) *Woman-Southern-Davis*[3] which describes a set of Southern women and the events they attended) [8]. Table 1 briefly summarizes these datasets.

Table 1. A brief description of tested datasets.

Dataset	No. objs	No. attrs	Max. intent	Lattice size
Mushroom	8416	128	22	238,710
Phytotherapy	142	108	11	304
WomenDavis	18	14	8	67

To evaluate our proposed LDS algorithm, we compared the results of the following five tested algorithms:

- **Sobol**: LDS algorithm based on Sobol sequence.
- **Scrambled VDC-2**: LDS algorithm based on Scrambled Van der corput sequence with radical base 2.
- **Scrambled VDC-3**: LDS algorithm based on Scrambled Van der corput sequence with radical base 3.
- **Scrambled VDC-5**: LDS algorithm based on Scrambled Van der corput sequence with radical base 5.
- **Monte Carlo**: MCS algorithm — which can serve as a good baseline here.

To find robust answers to the proposed questions, we conducted our empirical study with the chosen algorithms under the following setting:

- Vary the number of samples N from 1 up to 100. This could be helpful to precisely judge the convergence rate.

---

[3] Publicly available: http://fimi.ua.ac.be/data/.

- Re-run Monte Carlo algorithm for a maximum number of 1000 iterations. This is important to properly assess Monte Carlo which is purely random and sensitive to the starting point.
- Consider the concepts with the maximum intent sizes in order to asses how LDS can perform in extreme (or large) cases.

We then assess the accuracy and scalability of results by recording both the estimated stability and the elapsed time obtained at each sample size. These results are then used to calculate the following two metrics over intent sizes:

(1) The average absolute error ξ_a of estimated stability, which is computed as:

$$\xi_a = \frac{1}{|\mathcal{C}|} \sum_{c_k \in \mathcal{C}} |\tilde{\sigma}(c_k) - \sigma(c_k)| \tag{10}$$

where $\mathcal{C}$ is the set of concepts with the same intent size a, and $|\mathcal{C}|$ is its cardinality. $\tilde{\sigma}(c_k)$ is the estimated stability of the concept c_k and $\sigma(c_k)$ is the exact value of stability.

(2) Average elapsed time τ_a, which is computed as:

$$\tau_a = \frac{1}{|\mathcal{C}|} \sum_{c_k \in \mathcal{C}} t_k \tag{11}$$

where t_k is the elapsed time for approximating the stability of concept c_k.

All the experiments were run on an Intel(R) Core(TM) i7-2600 CPU @ 3.20 GHz computer with 16 GB of memory under Windows 10. We implemented our LDS algorithm as an extension to the Concepts 0.7.11 package that is implemented by Sebastian Bank.[4] For generating low-discrepancy sequences, we used SOBOL library.[5]

4.2 Results and Discussion

In terms of accuracy and performance, the results in Fig. 1 illustrate that MCS is always less accurate than all the tested variants of LDS, including Sobol and Scrambled VDC at bases 2, 3 and 5 on large data sets such as mushroom with large-sized intents. The same conclusion applies for medium and small datasets with concepts of small-sized intents such as Phytotherapy with intent size 3 (see Fig. 2) and Woman-Southern-Davis with intent sizes 4, 5 and 7 (see Fig. 3).

In terms of convergence behavior, it is clear that Scrambled VDC-3 and Scrambled VDC-5 compete at different intent sizes, both of them achieve the most accurate convergence rate of all the algorithms compared on the three underlying datasets. Scrambled VDC-2 and Sobol have frequently a similar accuracy and stable convergence behavior on large-sized intents. This may be due

---

[4] Publicly available: https://pypi.python.org/pypi/concepts.
[5] Publicly available: http://people.sc.fsu.edu/~jburkardt/py_src/sobol/sobol.html.

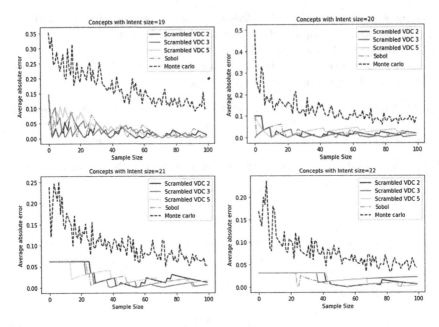

Fig. 1. The average absolute error ξ_a of LDS and MCS on *Mushroom* with intent size $a \in \{19, 20, 21, 22\}$. Each plot is labelled with the size of intent.

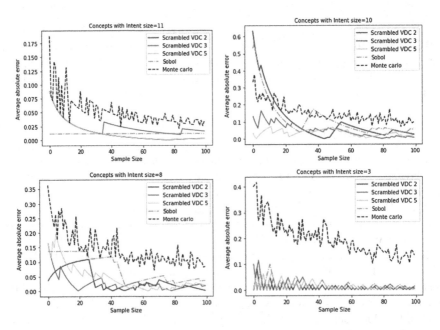

Fig. 2. The average absolute error ξ_a of LDS and MCS on *Phytotherapy* with intent size $a \in \{3, 8, 10, 11\}$. Each plot is labelled with the size of intent.

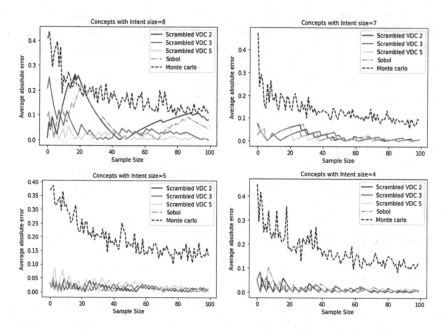

Fig. 3. The average absolute error ξ_a of LDS and MCS on *Woman-Southern-Davis* with intent size $a \in \{4, 5, 7, 8\}$. Each plot is labelled with the size of intent.

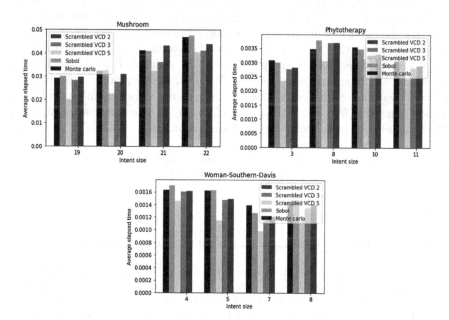

Fig. 4. The average elapsed time τ_a (in mins) for LDS and MCS on *Mushroom, Phytotherapy* and *Woman-Southern-Davis*.

to the fact that both of them use a binary representation (i.e., base 2) in the sequence generation process. MCS has often an unstable convergence rate when the sample size is less than 90, which could explain its need to increase the sample size for a better convergence behavior.

In terms of computational time, the results in Fig. 4 show that Scrambled VDC-5 dominates all other tested algorithms on all intent sizes. This is due to the fact that Scrambled VDC generates its sequence points using radical base 5 which requires less computational time than base 2 or 3. Apart from Scrambled VDC-5, none of the tested algorithms—the Sobol, the scrambled VDC at bases 2 and 3 and MCS — shows its clear superiority over the others.

Overall, the results in Figs. 1, 2, 3 and 4 clearly show that LDS outperforms MCS for approximating stability on all underlying datasets. Under the same number of samples, LDS can converge to more accurate estimations than MCS in the presence of large-sized concepts, and even with less computational time using some LDS variants (e.g., Scrambled VDC-5).

5   Conclusion

We have proposed LDS, a two-stage method that exploits LDS to efficiently estimate the stability index. To improve the accuracy of sampling, low-discrepancy (or quasi-random) sequences are first generated in order to eliminate the clumping of the sampled subsets and allow the uniform selection of correlated subsets across all the areas of the intent (or extent) power set space. We focus here on using Scrambled Van der Corput and Sobol sequences but the LDS method is applicable to other low-discrepancy sequences in general. Experiments on real-life datasets show that LDS is able to greatly bound the convergence rate by several orders of magnitude compared to MCS, which is the commonly used method in FCA to approximate stability. In the future we plan to integrate LDS with other variance reduction techniques such as importance sampling and stratification sampling. We also intend to perform an on-line sampling scenario of LDS-based algorithm to tackle stream formal contexts.

Acknowledgment. We acknowledge the financial support of the Natural Sciences and Engineering Research Council of Canada (NSERC).

References

1. Alpen, É.: Précis de Phytotérapie. Édition Alpen (2010). www.alpen.mc/precis-de-phytotherapie
2. Babin, M.A., Kuznetsov, S.O.: Approximating concept stability. In: Domenach, F., Ignatov, D.I., Poelmans, J. (eds.) ICFCA 2012. LNCS (LNAI), vol. 7278, pp. 7–15. Springer, Heidelberg (2012). https://doi.org/10.1007/978-3-642-29892-9_7
3. Bache, K., Lichman, M.: Mushroom data set (2013). http://archive.ics.uci.edu/ml
4. Belohlavek, R., Macko, J.: Selecting important concepts using weights. In: Valtchev, P., Jäschke, R. (eds.) ICFCA 2011. LNCS (LNAI), vol. 6628, pp. 65–80. Springer, Heidelberg (2011). https://doi.org/10.1007/978-3-642-20514-9_7

5. Buzmakov, A., Kuznetsov, S.O., Napoli, A.: Is concept stability a measure for pattern selection? Procedia Comput. Sci. **31**, 918–927 (2014)
6. Buzmakov, A., Kuznetsov, S.O., Napoli, A.: Scalable estimates of concept stability. In: Glodeanu, C.V., Kaytoue, M., Sacarea, C. (eds.) ICFCA 2014. LNCS (LNAI), vol. 8478, pp. 157–172. Springer, Cham (2014). https://doi.org/10.1007/978-3-319-07248-7_12
7. Caflisch, R.E.: Monte Carlo and quasi-Monte Carlo methods. Acta Numer. **7**, 1–49 (1998)
8. Davis, A., Gardner, B., Gardner, M.: Deep South (1941). http://networkdata.ics.uci.edu/netdata/html/davis.html
9. Faure, H., Tezuka, S.: Another random scrambling of digital (t, s)-sequences. In: Fang, K.T., Niederreiter, H., Hickernell, F.J. (eds.) Monte Carlo and Quasi-Monte Carlo Methods 2000, pp. 242–256. Springer, Heidelberg (2002). https://doi.org/10.1007/978-3-642-56046-0_16
10. Ganter, B., Wille, R.: Formal Concept Analysis: Mathematical Foundations. Springer, New York (1999). https://doi.org/10.1007/978-3-642-59830-2. Translator C. Franzke
11. Klimushkin, M., Obiedkov, S., Roth, C.: Approaches to the selection of relevant concepts in the case of noisy data. In: Kwuida, L., Sertkaya, B. (eds.) ICFCA 2010. LNCS (LNAI), vol. 5986, pp. 255–266. Springer, Heidelberg (2010). https://doi.org/10.1007/978-3-642-11928-6_18
12. Kuipers, L., Niederreiter, H.: Uniform distribution of sequences. Courier Corporation (2012)
13. Kuznetsov, S., Obiedkov, S., Roth, C.: Reducing the representation complexity of lattice-based taxonomies. In: Priss, U., Polovina, S., Hill, R. (eds.) ICCS-ConceptStruct 2007. LNCS (LNAI), vol. 4604, pp. 241–254. Springer, Heidelberg (2007). https://doi.org/10.1007/978-3-540-73681-3_18
14. Kuznetsov, S.O.: Learning of simple conceptual graphs from positive and negative examples. In: Żytkow, J.M., Rauch, J. (eds.) PKDD 1999. LNCS (LNAI), vol. 1704, pp. 384–391. Springer, Heidelberg (1999). https://doi.org/10.1007/978-3-540-48247-5_47
15. Kuznetsov, S.O.: On stability of a formal concept. Ann. Math. Artif. Intell. **49**(1), 101–115 (2007)
16. Kuznetsov, S.O., Makhalova, T.P.: Concept interestingness measures: a comparative study. In: Proceedings of the Twelfth International Conference on Concept Lattices and Their Applications, Clermont-Ferrand, France, 13–16 October 2015, pp. 59–72 (2015)
17. Kuznetsov, S.O., Makhalova, T.P.: On interestingness measures of formal concepts. CoRR abs/1611.02646 (2016)
18. Landau, D.P., Binder, K.: A Guide to Monte Carlo Simulations in Statistical Physics. Cambridge University Press, Cambridge (2014)
19. Lemieux, C.: Monte Carlo and quasi-Monte Carlo sampling (2009)
20. Muangprathub, J.: A novel algorithm for building concept lattice. Appl. Math. Sci. **8**(11), 507–515 (2014)
21. Niederreiter, H.: Random Number Generation and Quasi-Monte Carlo Methods. SIAM, Philadelphia (1992)
22. Owen, A.B.: Monte Carlo extension of quasi-Monte Carlo. In: Simulation Conference Proceedings Winter, vol. 1, pp. 571–577. IEEE (1998)
23. Roth, C., Obiedkov, S., Kourie, D.G.: On succinct representation of knowledge community taxonomies with formal concept analysis. Int. J. Found. Comput. Sci. **19**(02), 383–404 (2008)

24. Schretter, C., He, Z., Gerber, M., Chopin, N., Niederreiter, H.: Van der corput and golden ratio sequences along the hilbert space-filling curve. In: Cools, R., Nuyens, D. (eds.) Monte Carlo and Quasi-Monte Carlo Methods, vol. 163, pp. 531–544. Springer, Cham (2016). https://doi.org/10.1007/978-3-319-33507-0_28
25. Zhi, H.L.: On the calculation of formal concept stability. J. Appl. Math. **2014**, 1–6 (2014)

Lifted Most Probable Explanation

Tanya Braun$^{(\boxtimes)}$ and Ralf Möller

Institute of Information Systems, Universität zu Lübeck, Lübeck, Germany
{braun,moeller}@ifis.uni-luebeck.de

Abstract. Standard approaches for inference in probabilistic formalisms with first-order constructs include lifted variable elimination (LVE) for single queries, boiling down to computing marginal distributions. To handle multiple queries efficiently, the lifted junction tree algorithm (LJT) uses a first-order cluster representation of a knowledge base and LVE in its computations. Another type of query asks for a most probable explanation (MPE) for given events. The purpose of this paper is twofold: (i) We formalise how to compute an *MPE* in a lifted way with LVE and LJT. (ii) We present a *case study* in the area of IT security for risk analysis. A lifted computation of MPEs exploits symmetries, while providing a correct and exact result equivalent to one computed on ground level.

Keywords: Probabilistic logical model · Lifting · MPE · MAP
Abduction

1 Introduction

In recent years, IT security has become a major research area to perform tasks such as intrusion detection, risk analysis, or risk mitigation. These tasks need efficient and exact inference algorithms for large probabilistic models given sets of events and a multitude of queries to answer. Lifting uses symmetries in a model to speed up reasoning with known domain objects. Symmetries are bound to appear in networks at risk (e.g., several access points, user groups).

For single queries, researchers have sped up runtimes for inference significantly. Variable elimination (VE) decomposes a propositional model into subproblems to evaluate them in an efficient order [20], eliminating random variables (randvars) not present in a query. Lifted VE (LVE), also called FOVE, introduced in [12] and expanded in [8,13,19] to its current form GC-FOVE, exploits symmetries. For multiple queries, Lauritzen and Spiegelhalter present junction trees (jtrees), to represent clusters of randvars in a model, along with a reasoning algorithm [7]. In [1], we present the lifted junction tree algorithm (LJT) based on [7,19], using a first-order jtree (FO jtree). LJT imposes some static overhead but has a significant speed up compared to LVE for query answering (QA).

So far, queries concern marginal and conditional probability distributions. Another important inference task is to compute most probable explanations

P. Chapman et al. (Eds.): ICCS 2018, LNAI 10872, pp. 39–54, 2018.
https://doi.org/10.1007/978-3-319-91379-7_4

(MPEs), also known as abduction: We look for the most probable assignment to all randvars in a model. Pearl introduces the idea of MPEs and a propagation algorithm for singly-connected networks [11]. Dawid presents an algorithm to compute MPEs on jtrees [4], which allows for computing up to $k = 3$ MPEs [10]. Dechter formalises computing MPEs as a form of VE, replacing sums with max-out operations [5]. In [14], de Salvo Braz et al. adapt FOVE for MPEs, which does not incorporate GC-FOVE's advances. A related concept to MPE is maximum a posteriori assignment (MAP) for a most probable assignment to a subset of model randvars. Computing exact MAP solutions is intractable [5].

This paper extends LVE and LJT transferring the ideas of Dawid and Dechter, contributing the following. (i) We formalise how to compute an *MPE* with LVE and LJT and discuss MAPs and sets of queries. (ii) We present a *case study* in the area of IT security, specifically, on risk analysis, which our formalism easily models with probable assignments and likelihood of observations as common queries. LJT offers the advantage of clusters for local computations, identifying MAPs easily computable, and combining queries of different types easily.

Various research areas cover MPEs, from probabilistic logic programs [17] to probabilistic databases (DBs) [2]. Ceylan et al. show the complexity of computing most probable DBs and hypotheses jumping off Gribkoff et al.'s work on most probable DBs [6]. LJT offers compiling a DB into a compact model, even allowing for more expressiveness, for fast online QA w.r.t. varying query types. Schröder et al. study most probable state sequences in an agent scenario using lifting [15]. Muñoz-González et al. investigate exact inference for IT security, namely, for attack graphs (AGs), with a jtree algorithm performing best [9]. Similar scenarios appear, e.g., in healthcare [3].

The remainder of this paper is structured as follows: We introduce basic notations and recap LVE and LJT. Then, we present LVE and LJT for computing MPEs, followed by a discussion and a case study. We conclude with future work.

2   Preliminaries

This section introduces basic notations for models as well as the different query types and recaps LVE and LJT.

2.1   Parameterised Models

Parameterised models compactly represent propositional models with logical variables (logvars) as parameters in randvars. We set up a model for risk analysis with an AG. An AG models attacks and their targeted components in a network. A binary randvar holds if a component is compromised, which provides an attacker with privileges to further compromise a network towards a final target. Logvars represent users with certain privileges. We first denote basic blocks.

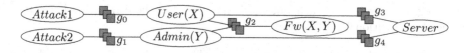

Fig. 1. Parfactor graph for G_{ex}

Definition 1. *Let* **L** *be a set of logvar names,* Φ *a set of factor names, and* **R** *a set of randvar names. A parameterised randvar (PRV)* $R(L_1, \ldots, L_n), n \geq 0$, *is a syntactical construct with a randvar* $R \in$ **R** *and logvars* $L_1, \ldots, L_n \in$ **L** *to represent a set of randvars. For PRV A, the term* $range(A)$ *denotes possible values. A logvar L has a domain, denoted* $\mathcal{D}(L)$. *A constraint* $(\mathbf{X}, C_{\mathbf{X}})$ *is a tuple with a sequence of logvars* $\mathbf{X} = (X_1, \ldots, X_n)$ *and a set* $C_{\mathbf{X}} \subseteq \times_{i=1}^{n} \mathcal{D}(X_i)$ *restricting logvars to certain values. The symbol* $\top$ *marks that no restrictions apply and may be omitted. The term* $lv(P)$ *refers to the logvars in some* P, $rv(P)$ *to the PRVs with constraints, and* $gr(P)$ *denotes the set of instances of* P *with its logvars grounded w.r.t. constraints.* P *may be PRV, parfactor, or model.*

From **R** $= \{Server, User\}$ and **L** $= \{X, Y\}$ with $\mathcal{D}(X) = \{x_1, x_2, x_3\}$ and $\mathcal{D}(Y) = \{y_1, y_2\}$, we build the binary PRVs $Server$ and $User(X)$. With $C = (X, \{x_1, x_2\})$, $gr(User(X)|C) = \{User(x_1), User(x_2)\}$. $gr(User(X)|\top)$ also contains $User(x_3)$. Next, parametric factors (parfactors) combine PRVs.

Definition 2. *A parfactor* g *describes a function, mapping argument values to real values, i.e., potentials. We denote a parfactor by* $\forall \mathbf{X} : \phi(\mathcal{A}) \mid C$ *where* $\mathbf{X} \subseteq$ **L** *is a set of logvars.* $\mathcal{A} = (A_1, \ldots, A_n)$ *is a sequence of PRVs, each built from* **R** *and possibly* **X**. *We omit* $(\forall \mathbf{X} :)$ *if* $\mathbf{X} = lv(\mathcal{A})$. $\phi : \times_{i=1}^{n} range(A_i) \mapsto \mathbb{R}^+$ *is a function with name* $\phi \in \Phi$. ϕ *is identical for all groundings of* $\mathcal{A}$. C *is a constraint on* **L**. *A set of parfactors forms a* model $G := \{g_i\}_{i=1}^{n}$. G *represents the probability distribution* $P_G = \frac{1}{Z} \prod_{f \in gr(G)} f$, *where* Z *is the normalisation constant.*

We also build the binary PRVs $Attack1$, $Attack2$, $Admin(Y)$, and $Fw(X,Y)$ to model two different attacks, an admin user with more privileges than a regular user $User(X)$, and a firewall between admin and regular users. Model G_{ex} reads $\{g_i\}_{i=0}^{4}$, $g_0 = \phi_0(Attack1, User(X))$, $g_1 = \phi_1(Attack2, Admin(Y))$, $g_2 = \phi_2(User(X), Admin(Y), Fw(X,Y))$, $g_3 = \phi_3(Server, User(X))$, $g_4 = \phi_4(Server, Admin(Y))$. g_2 has eight, the others four input-output pairs (omitted). Constraints are $\top$, e.g., $gr(g_0)$ contains three factors with identical ϕ_0. Figure 1 depicts G_{ex} as a graph with six variable nodes for the PRVs and five factor nodes for g_0 to g_4 with edges to the PRVs involved.

A counting randvar (CRV) encodes for n interchangeable randvars how many have a certain value, exploiting symmetries in potentials. In $\phi(U_1, U_2, U_3)$ as below left, the potentials are the same for two U_i being true ($= 1$) and one false ($= 0$) or vice versa. Using a logvar N, a CRV, denoted as $\#_N[U(N)]$, and a histogram as range value, the mapping on the right carries the same information.

A histogram specifies for each value of U how many of the n randvars have this value (first position $U = 1$, second $U = 0$).

$$(0,0,0) \rightarrow 1, (0,0,1) \rightarrow 2, (0,1,0) \rightarrow 2, (0,1,1) \rightarrow 3, \quad [0,3] \rightarrow 1, [1,2] \rightarrow 2,$$
$$(1,0,0) \rightarrow 2, (1,0,1) \rightarrow 3, (1,1,0) \rightarrow 3, (1,1,1) \rightarrow 2 \quad [2,1] \rightarrow 3, [3,0] \rightarrow 2$$

Definition 3. $\#_X[P(\mathbf{X})]$ *denotes a CRV with PRV* $P(\mathbf{X})$, *where* $lv(\mathbf{X}) = \{X\}$. *Its range is the space of possible histograms. If* $\{X\} \subset lv(\mathbf{X})$, *the CRV is a* parameterised CRV (PCRV) *representing a set of CRVs. Since counting binds logvar* X, $lv(\#_X[P(\mathbf{X})]) = \mathbf{X} \setminus \{X\}$. *A histogram* h *is a set of pairs* $\{(v_i, n_i)\}_{i=1}^m$, $v_i \in range(P(\mathbf{X}))$, $m = |range(P(\mathbf{X}))|$, $n_i \in \mathbb{N}$, *and* $\sum_i n_i = |gr(P(\mathbf{X})|C)|$. *A shorthand notation is* $[n_1, \ldots, n_m]$. $h(v_i)$ *returns* n_i. *Adding two histograms of a PRV yields* $\{(v_i, n_i + n_i')\}_{i=1}^m$, *multiplying one with a value* c *yields* $\{(v_i, c \cdot n_i)\}_{i=1}^m$.

The *semantics* of a model is given by grounding and building a full joint distribution. QA asks for a likelihood of an event, a marginal distribution of a set of randvars, or a conditional distribution given events, all types boiling down to computing marginals w.r.t. a model's joint distribution. Formally, $P(\mathbf{Q}|\mathbf{E})$ denotes a query with $\mathbf{Q}$ a set of grounded PRVs and $\mathbf{E}$ a set of events (grounded PRVs with range values). For G_{ex}, $P(Admin(y_1)|Server = 1)$ asks for the conditional distribution of $Admin(y_1)$ with $Server = 1$ (compromised) as an event.

MPE and MAP are assignment queries. An MPE for a model G and evidence $\mathbf{E}$ is the most probable assignment to all remaining randvars, i.e., $mpe(G, \mathbf{E}) = \arg\max_{\mathbf{a} \in range(rv(G) \setminus rv(\mathbf{E}))} P(\mathbf{a}|\mathbf{E})$. A MAP for a set of randvars $\mathbf{V}$ is the most probable assignment to $\mathbf{V}$ given $\mathbf{E}$ summing over the remaining randvars, i.e., $map(G, \mathbf{V}, \mathbf{E}) = \arg\max_{\mathbf{v} \in range(\mathbf{V})} \sum_{\mathbf{a}} P(\mathbf{v}, \mathbf{a}|\mathbf{E})$, $\mathbf{a} \in range(rv(G) \setminus \mathbf{V} \setminus rv(\mathbf{E}))$. While MPE is relatively easy to compute, with $\arg\max$ operations only, MAP is not as it involves sums and $\arg\max$ operations, which are not commutative.

2.2 Query Answering Algorithms

LVE and LJT answer queries for probability distributions. LJT uses an FO jtree with LVE as a subroutine. We briefly recap LVE and LJT.

Lifted Variable Elimination: LVE exploits symmetries that lead to duplicate calculations. In essence, it computes VE for one case and exponentiates its result for isomorphic instances (lifted summing out). GC-FOVE implements LVE through an operator suite (cf. [19] for details). Its main operator *sum-out* realises lifted summing out. An operator *absorb* handles evidence in a lifted way. The remaining operators (*count-convert, split, expand, count-normalise, multiply, ground-logvar*) aim at enabling a lifted summing out, transforming part of a model. All operators have pre- and postconditions to ensure computing a result equivalent to one computed on $gr(G)$. Algorithm 1 shows an outline. To answer a query, LVE eliminates all non-query randvars. For a new query, LVE starts over.

Algorithm 1. Outline of LVE	**Algorithm 2.** Outline of LJT
LVE(Model G, Query $\mathbf{Q}$, Evidence $\mathbf{E}$)	LJT(Model G, Queries $\{\mathbf{Q}_i\}_{i=1}^{m}$, Evid. $\mathbf{E}$)
Absorb $\mathbf{E}$ in G	Construct FO jtree J
while G has non-query PRVs **do**	Enter $\mathbf{E}$ into J
if PRV A fulfils *sum-out* prec. **then**	Pass messages on J
Eliminate A using *sum-out*	**for** each query $\mathbf{Q}_i$ **do**
else	Find subtree J_i for $\mathbf{Q}_i$
Apply transformator	Extract submodel G_i from J_i
return MULTIPLY(G) ▷ normalise	LVE(G_i, $\mathbf{Q}_i$, ∅)

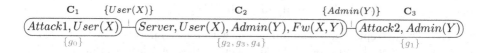

Fig. 2. FO jtree for G_{ex} (local parcluster models in grey)

Lifted Junction Tree Algorithm: Algorithm 2 outlines LJT for a set of queries. LJT first constructs an FO jtree with parameterised clusters (parclusters) as nodes, which are sets of PRVs connected by parfactors, defined as follows.

Definition 4. *An FO jtree for a model G is a cycle-free graph $J = (V, E)$, where V is the set of nodes and E the set of edges. Each node in V is a parcluster $\mathbf{C}_i$. A parcluster is denoted by $\forall \mathbf{L} : \mathcal{A} \mid C$. $\mathbf{L}$ is a set of logvars. $\mathcal{A}$ is a set of PRVs with $lv(\mathcal{A}) \subseteq \mathbf{L}$. We omit $(\forall \mathbf{L} :)$ if $\mathbf{L} = lv(\mathcal{A})$. Constraint C restricts $\mathbf{L}$. $\mathbf{C}_i$ has a local model G_i of assigned parfactors. An assigned parfactor $\phi(\mathcal{A}_\phi) \mid C_\phi$ must fulfil (i) $\mathcal{A}_\phi \subseteq \mathcal{A}$, (ii) $lv(\mathcal{A}_\phi) \subseteq \mathbf{L}$, and (iii) $C_\phi \subseteq C$.*
An FO jtree must satisfy three properties: (i) A parcluster $\mathbf{C}_i$ is a set of PRVs from G. (ii) For every parfactor $g = \phi(\mathcal{A}) \mid C$ in G, $\mathcal{A}$ appears in some $\mathbf{C}_i$. (iii) If a PRV from G appears in $\mathbf{C}_i$ and $\mathbf{C}_j$, it must appear in every parcluster on the path between nodes i and j in $\mathcal{J}$. The parameterised set $\mathbf{S}_{ij}$, called separator of edge i—j in $\mathcal{J}$, contains the shared randvars of $\mathbf{C}_i$ and $\mathbf{C}_j$, i.e., $\mathbf{C}_i \cap \mathbf{C}_j$.

For G_{ex}, Fig. 2 depicts an FO jtree with three parclusters, $\mathbf{C}_1 = \forall X: \{Attack1, User(X)\} \mid \top$, $\mathbf{C}_2 = \forall X, Y: \{Server, User(X), Admin(Y), Fw(X, Y)\} \mid \top$, $\mathbf{C}_3 = \forall Y: \{Attack2, Admin(Y)\} \mid \top$. Separators are $\mathbf{S}_{12} = \mathbf{S}_{21} = \{User(X)\}$ and $\mathbf{S}_{23} = \mathbf{S}_{32} = \{Admin(Y)\}$. Each parcluster has a non-empty local model.

After constructing an FO jtree, LJT enters evidence for the local models to absorb it. Message passing propagates local information through the FO jtree in two passes: First, LJT sends messages from the periphery towards the center. Second, LJT sends messages the opposite way. A message is a set of parfactors over separator PRVs. For a message from node i to neighbour j, LJT eliminates all PRVs not in separator $\mathbf{S}_{ij}$ from G_i and the messages from other neighbours using LVE. Afterwards, each parcluster holds all information of the model in its local model and messages. LJT answers a query by finding a subtree whose parclusters cover the query randvars, extracting a submodel of local models and messages, and using LVE to answer the query on the submodel. Next, we adapt LVE in the form of GC-FOVE to compute an MPE.

3   Lifted Algorithms for Most Probable Explanations

We adapt LVE and LJT to compute MPEs. We use the GC-FOVE operator suite as a basis for LVE^{mpe}, which becomes a subroutine in LJT^{mpe}.

3.1   Most Probable Explanation with LVE

For LVE^{mpe}, we replace lifted summing out with a lifted maxing out that picks a maximum potential and retains the argument values that lead to that potential for a set of interchangeable randvars, extending parfactors as a consequence, and adapt GC-FOVE. At the end, we look at G_{ex} to compute an MPE. We assume familiarity with operations from relational algebra such as renaming ρ, join $\bowtie$, selection σ, and projection π. We need a count function, the notion of count normalisation, and alignments for the operators.

Definition 5. *Given a constraint $C = (\mathbf{X}, C_\mathbf{X})$, for any $\mathbf{Y} \subseteq \mathbf{X}$ and $\mathbf{Z} \subseteq \mathbf{X} \setminus \mathbf{Y}$, the function $\text{COUNT}_{\mathbf{Y}|\mathbf{Z}} : C_\mathbf{X} \to \mathbb{N}$ tells us how many values of $\mathbf{Y}$ co-occur with the value of $\mathbf{Z}$ in $C_\mathbf{X}$, defined by $\text{COUNT}_{\mathbf{Y}|\mathbf{Z}}(t) = |\pi_\mathbf{Y}(\sigma_{\mathbf{Z} = \pi_\mathbf{Z}(t)}(C_\mathbf{X}))|$. We define $\text{COUNT}_{\mathbf{Y}|\mathbf{Z}}(t) = 1$ for $\mathbf{Y} = \emptyset$. $\mathbf{Y}$ is count-normalised w.r.t. $\mathbf{Z}$ in C iff $\exists n \in \mathbb{N}$ s.t. $\forall t \in C_\mathbf{X} : \text{COUNT}_{\mathbf{Y}|\mathbf{Z}}(t) = n$, denoted by $\text{COUNT}_{\mathbf{Y}|\mathbf{Z}}(C)$.*

Definition 6. *A substitution $\theta = \{X_i \to t_i\}_{i=1}^n = \{\mathbf{X} \to \mathbf{t}\}$ replaces each occurrence of logvar X_i with term t_i (a logvar or domain value). An alignment θ between two parfactors $\phi_1(\mathcal{A}_1)|C_1$ and $\phi_2(\mathcal{A}_2)|C_2$ is a one-to-one substitution $\{\mathbf{X}_1 \to \mathbf{X}_2\}$ with $\mathbf{X}_1 \subseteq lv(\mathcal{A}_1)$ and $\mathbf{X}_2 \subseteq lv(\mathcal{A}_2)$ s.t. $\rho_\theta(\pi_{\mathbf{X}_1}(C_1)) = \pi_{\mathbf{X}_2}(C_2)$.*

Operator 1 defines *max-out* to replace *sum-out*. The inputs (parfactor g, PRV A_i) and preconditions are identical. The composition of output ϕ' and the postcondition differ. To eliminate A_i, *max-out* selects for each valuation $(\dots, a_{i-1}, a_{i+1}, \dots)$ the maximum potential p_i in $range(A_i)$ and its corresponding arg max value a. The new ϕ' maps $(\dots, a_{i-1}, a_{i+1}, \dots)$ to p_i exponentiated with $r = \text{COUNT}_{\mathbf{X}^{excl}|\mathbf{X}^{com}}(C)$. r denotes the number of instances that the logvars only appearing in A_i ($\mathbf{X}^{excl}$) stand for w.r.t. the remaining logvars ($\mathbf{X}^{com}$).

To retain the range value a, we map arguments in parfactors not only to a potential but also to range values for maxed out PRVs. We store the range values in histograms to have an identical representation for both maxed out PRVs and PCRVs. Histograms also enable us to encode how many instances $\mathbf{X}^{excl}$ represent so as to not carry around constraints for eliminated logvars. Thus, function ϕ in a parfactor $\phi(\mathcal{A})|C$ maps arguments to a pair of a potential and a set of histograms $\{h_{V_1}, \dots, h_{V_l}\} = \bigcup_{j=1}^l h_{V_j}$ for already maxed out PRVs V_j.

Operator 1. Lifted Maxing-out

Operator MAX-OUT
Inputs:
(1) $g = \phi(\mathcal{A})|C$: a parfactor in G
(2) A_i: an atom in $\mathcal{A}$, to be summed out from g
Preconditions:
(1) For all PRVs $\mathcal{V}$, other than A_i, in G: $rv(\mathcal{V}) \cap rv(A_i|C) = \emptyset$
(2) A_i contains all the logvars $X \in lv(\mathcal{A})$ for which $\pi_X(C)$ is not singleton
(3) $\mathbf{X}^{excl} = lv(A_i) \setminus lv(\mathcal{A}\setminus A_i)$ count-normalised w.r.t. $\mathbf{X}^{com} = lv(A_i) \cap lv(\mathcal{A}\setminus A_i)$ in C
Output: $\phi'(\mathcal{A}')|C'$, such that
(1) $\mathcal{A}' = \mathcal{A}\setminus A_i$
(2) $C' = \pi_{\mathbf{X}^{com}}(C)$
(3) for each valuation $\mathbf{a}' = (\ldots, a_{i-1}, a_{i+1}, \ldots)$ of $\mathcal{A}'$,
given $\phi(\ldots, a_{i-1}, a_i, a_{i+1}, \ldots) = (p_i, \bigcup_{j=1}^{l} h_{i,V_j})$ and $r = \mathrm{COUNT}_{\mathbf{X}^{excl}|\mathbf{X}^{com}}(C)$,
$\phi'(\mathbf{a}') = \left(p', \{h_{A_i}\} \cup \bigcup_{j=1}^{l} h'_{i,V_j}\right)$,
$p' = (\max_{a_i \in range(A_i)} p_i)^r$,
$h_{A_i} = r \cdot h$, $h = \arg\max_{a_i \in range(A_i)} p_i$ if A_i PCRV,
$h_{A_i} = \{(a_i, n_i)\}_{i=1}^{|range(A_i)|}$ otherwise.
$n_i = r$ for $a_i = \arg\max_{a_i \in range(A_i)} p_i$,
$n_i = 0$ otherwise.
$h'_{i,V_j} = c_j \cdot h_{i,V_j}$, $c_j = \mathrm{COUNT}_{\mathbf{X}^{excl} \cap lv(V_j)|\mathbf{X}^{com}}(C)$
Postcondition: $mpe(G \setminus \{g\} \cup \{\mathrm{MAX\text{-}OUT}(g, A_i)\}) = \arg\max_{rv(A_i|C)} G$

The logvars in $\mathbf{X}^{excl}$ disappear from the arguments of g with the elimination of A_i and they need to be accounted for in the histograms as well: If A_i is a PRV, assignment a is stored in a peak-shaped histogram h_{A_i} where a maps to r and the other values to 0. If A_i is a (P)CRV, a already is a histogram h to multiply with r, i.e., $h_{A_i} = r \cdot h$. The histograms are multiplied with $\mathrm{COUNT}_{\mathbf{X}^{excl} \cap lv(V_j)|\mathbf{X}^{com}}(C)$, to account for logvars in $\mathbf{X}^{excl}$ appearing in any maxed out PRV V_j. h_{A_i} is added to the set of histograms. The postcondition reflects the max-out operation.

The operator *absorb* is only applied at the beginning to absorb evidence. The sets of histograms are empty. The operators *split*, *expand*, *count-normalise*, and *ground-logvar* all involve duplicating a parfactor and partitioning constraints. The operators *split* and *expand*, which implement splitting operations for PRVs and CRVs respectively, do so s.t. sets of randvars do not overlap (ensure first precondition of *max-out*). The operator *count-normalise* duplicate and partition to count-normalise a set of logvars w.r.t. another set (ensure third precondition of *max out*). Operator *ground-logvar* implements grounding through duplication and partitioning as well. Duplicating a parfactor also duplicates its histograms. Assignments are unaffected as we only change constraints. If a constraint changes for a logvar that appears in a maxed out PRV, the change also applies to it.

The operators *multiply* and *count-convert* remain. Operator 2 shows how to multiply two parfactors. Their histograms concern different PRVs, else LVE would have multiplied them earlier. For each new input, histogram sets unify.

Operator 3. Count Conversion

Operator COUNT-CONVERT
Inputs:
(1) $g = \phi(\mathcal{A})|C$: a parfactor in G
(2) X: a logvar in $lv(\mathcal{A})$
Preconditions:
(1) there is exactly one atom $A_i \in \mathcal{A}$ with $X \in lv(A_i)$
(2) X is count-normalised w.r.t. $lv(\mathcal{A}) \setminus \{X\}$ in C
(3) for all counted logvars $X^\#$ in g: $\pi_{X,X^\#}(C) = \pi_X(C) \times \pi_{X^\#}(C)$
Output: $\phi'(\mathcal{A}')|C$, such that
(1) $\mathcal{A}' = \mathcal{A} \setminus A_i \cup A_i'$ with $A_i' = \#_X[A_i]$
(2) for each valuation $\mathbf{a}'$ to $\mathcal{A}'$ with $a_i' = h$: $\phi'(\ldots, a_{i-1}, h, a_{i+1}, \ldots) = \left(p', \bigcup_{j=1}^l h_{V_j}' \right)$,

given $\phi(\ldots, a_{i-1}, a_i, a_{i+1}, \ldots) = (p_i, \bigcup_{j=1}^l h_{i,V_j})$

$p' = \prod_{a_i \in range(A_i)} p_i^{h(a_i)}$, $h_{V_j}' = \sum_{a_i \in range(A_i)} h_{i,V_j}$ if $X \in lv(V_j)$, $h_{V_j}' = h_{V_j}$ oth.
Postcondition: $G \equiv G \setminus \{g\} \cup \{\text{COUNT-CONVERT}(g, X)\}$

Operator 2. Lifted Multiplication

Operator MULTIPLY
Inputs:
(1) $g_1 = \phi_1(\mathcal{A}_1)|C_1$: a parfactor in G
(2) $g_2 = \phi_2(\mathcal{A}_2)|C_2$: a parfactor in G
(3) $\theta = \{\mathbf{X}_1 \to \mathbf{X}_2\}$: an alignment between g_1 and g_2
Preconditions:
(1) for $i = 1, 2$: $\mathbf{Y}_i = lv(\mathcal{A}_i) \setminus \mathbf{X}_i$ is count-normalised w.r.t. $\mathbf{X}_i$ in C_i
Output: $\phi(\mathcal{A})|C$, with
(1) $\mathcal{A} = \mathcal{A}_1\theta \cup \mathcal{A}_2$, and
(2) $C = \rho_\theta(C_1) \bowtie C_2$
(3) for each valuation $\mathbf{a}$ of $\mathcal{A}$, with $\mathbf{a}_1 = \pi_{\mathcal{A}_1\theta}(\mathbf{a})$ and $\mathbf{a}_2 = \pi_{\mathcal{A}_2}(\mathbf{a})$:
given $\phi_1(\mathbf{a}_1) = (p_1, \bigcup_{j_1=1}^{l_1} h_{V_{j_1}})$, $\phi_2(\mathbf{a}_2) = (p_2, \bigcup_{j_2=1}^{l_2} h_{V_{j_2}})$,

$\phi(\mathbf{a}) = \left(p_1^{\frac{1}{r_2}} \cdot p_2^{\frac{1}{r_1}}, \bigcup_{j_1=1}^{l_1} h_{V_{j_1}} \cup \bigcup_{j_2=1}^{l_2} h_{V_{j_2}} \right)$, with $r_i = \text{COUNT}_{\mathbf{Y}_i|\mathbf{X}_i}(C_i)$
Postcondition: $G \equiv G \setminus \{g_1, g_2\} \cup \{\text{MULTIPLY}(g_1, g_2, \theta)\}$

Operator 3 shows a count conversion of a logvar X in a parfactor $g = \phi(\mathcal{A})|C$ yielding a (P)CRV. While no other PRV in g may contain X, maxed out PRVs may contain X. Each new valuation $(\ldots, a_{i-1}, h, a_{i+1}, \ldots)$, where h is a histogram for the PCRV, maps to a new potential calculated as before and a set of histograms that are updated. If a maxed out PRV V_j contains the newly counted logvar X, it makes the number of instances of X explicit in V_j: It adds the histograms mapped to by $\phi(\ldots, a_{i-1}, a_i, a_{i+1}, \ldots)$ $h(a_i)$ times for each a_i.

Let us take a look at an example for *max-out*, *multiply*, and *count-convert*. Consider parfactor $g_2 = \phi_2(User(X), Admin(Y), Fw(X, Y))$ in G_{ex}. PRV $Fw(X, Y)$ contains all logvars and does not appear further in G_{ex}. No logvars disappear after its elimination ($r = 1$). Assume the following mapping with random potentials:

$(0,0,0) \rightarrow (0.9, \emptyset)$, $(0,0,1) \rightarrow (0.1, \emptyset)$, $(0,1,0) \rightarrow (0.4, \emptyset), (0,1,1) \rightarrow (0.6, \emptyset)$,
$(1,0,0) \rightarrow (0.55, \emptyset), (1,0,1) \rightarrow (0.45, \emptyset), (1,1,0) \rightarrow (0.2, \emptyset), (1,1,1) \rightarrow (0.8, \emptyset)$

The new parfactor reads $g_2' = \phi_2'(User(X), Admin(Y))$. For input $(0,0)$, we build the output from $(0,0,0) \rightarrow 0.9$ and $(0,0,1) \rightarrow 0.1$. The maximum potential is 0.9 with assignment 0 for $Fw(X,Y)$, i.e., $(0,0) \rightarrow (0.9^1, [0,1])$. The other pairs are:

$$(0,1) \rightarrow (0.6^1, [1,0]), \quad (1,0) \rightarrow (0.55^1, [0,1]), \quad (1,1) \rightarrow (0.8^1, [1,0]).$$

To max out $User(X)$, we need to multiply g_0, g_2', and g_3, leading to a parfactor $g_m = \phi_m(Attack1, User(X), Admin(Y), Server)$. For new input $(0,0,0,0)$, the output is the product of the potentials for the potential part. For the histogram set, the output is the union of $\emptyset$, $[0,1]$, and $\emptyset$, to which $\phi_0(0,0)$, $\phi_2'(0,0)$, and $\phi_3(0,0)$ map. Next, we count-convert Y, yielding parfactor $g_\# = \phi_\#(Attack1, User(X), \#_Y[Admin(Y)], Server)$ with $range(\#_Y[Admin(Y)]) = \{[0,2], [1,1], [2,0]\}$. Since Y appears in $Fw(X,Y)$, we account for $|gr(Y|\top)| = 2$ Y values in the histograms for $Fw(X,Y)$. Given original pairs $\phi_m(0,0,0,0) \rightarrow (2, \{[0,1]\})$ and $\phi_m(0,0,1,0) \rightarrow (1, \{[1,0]\})$, new pairs are:

$$\phi_{12}'(0,0,[0,2],0) \rightarrow (1^0 \cdot 2^2, \{[0,1] + [0,1]\}) = (4, \{[0,2]\})$$
$$\phi_{12}'(0,0,[1,1],0) \rightarrow (1^1 \cdot 2^1, \{[1,0] + [0,1]\}) = (2, \{[1,1]\})$$
$$\phi_{12}'(0,0,[2,0],0) \rightarrow (1^2 \cdot 2^0, \{[1,0] + [1,0]\}) = (1, \{[2,0]\})$$

Now, we can max out $User(X)$ (includes multiplying the histograms for $Fw(X,Y)$ with $\text{COUNT}_{\{X\} \cap \{X,Y\}|\emptyset}(\top) = 3$). Next, we set up LVE^{mpe}.

LVE^{mpe}: Algorithm 3 outlines LVE^{mpe} with model G and evidence $\mathbf{E}$ as input. Having absorbed $\mathbf{E}$, it maxes out all PRVs in G, applying transformators if necessary. The result is a parfactor with no arguments, a potential, and histograms for each PRV in $rv(G) \setminus rv(\mathbf{E})$ stating how many instances have a specific value.

Consider an MPE for G_{ex} without evidence. LVE^{mpe} maxes out $Fw(X,Y)$ and $User(X)$ as given above, yielding $g_\#' = \phi_\#'(Attack1, \#_Y[Admin(Y)], Server)$. It multiplies g_1 and g_4, count-converts Y in the product, and multiplies the result into $g_\#'$. It then maxes out the CRV and the remaining randvars, which results in a parfactor with an empty argument mapping to a potential and a set of histograms, e.g., $\phi() \rightarrow (p, \{[0,6]_{Fw}, [0,3]_{Us}, [0,2]_{Ad}, [0,1]_{Se}, [0,1]_{At1}, [0,1]_{At2}\})$. Now, we argue why LVE^{mpe} is sound.

Theorem 1. *LVE^{mpe} is sound, i.e., computes an MPE for model G equivalent to an MPE computed for $gr(G)$.*

Proof. We assume that LVE as specified by GC-FOVE is correct. Thus, potentials are handled correctly. Replacing $\sum$ with $\arg\max$ produces a correct MPE in the ground case. For a set of interchangeable randvars, the $\arg\max$ assignment is identical given each possible valuation of the remaining randvars. So, assigning one value for all instances of a PRV, as in *max-out*, is correct. Storing the assignments in histograms based on counts is a different representation to avoid storing constraints for logvars only appearing in maxed out PRVs. The enabling operators, except count conversion, do not affect histograms of maxed out PRVs. When count-converting a logvar, we need to count the instances of the same logvar in maxed out PRVs accumulating assignments based on the histogram of the new CRV. With correct eliminations and count representation in histograms, LVE^{mpe} computes an MPE equivalent to one computed on a ground level.   □

3.2   Most Probable Explanation with LJT

We adapt LJT to compute an MPE by calculating messages using LVE^{mpe} to max out non-separator PRVs. Messages carry over the assignments of maxed out PRVs. As outlined in Algorithm 4, LJT^{mpe} constructs an FO jtree J for an input model G, enters evidence $\mathbf{E}$ into J, and passes messages in J, which only needs an inward pass. At the innermost node, it maxes out the remaining PRVs and returns an MPE. For new evidence, LJT^{mpe} starts over at entering evidence.

Computing an MPE for G_{ex} without evidence starts with constructing an FO jtree as seen in Fig. 2. Without evidence, message passing commences. Nodes 1 and 3 prepare a message for node 2 using LVE^{mpe}. At both nodes, the logvar is count-converted to max out the ground PRV leading to a message from node 1 over $\#_X(User(X))$ with an assignment for $Attack1$ and a message from node 3 over $\#_Y(Admin(Y))$ with an assignment for $Server$. Node 2 receives both messages and maxes out its PRVs to complete the MPE. After maxing out $Fw(X,Y)$ in g_2, it multiplies the result with g_3 and g_4 and count-converts X and Y to multiply each message in. Then, it maxes out $\#_X(User(X))$ and $\#_Y(Admin(Y))$ as well as the remaining randvars producing a parfactor with an empty argument that maps to a potential and a set of histograms identical to the one LVE^{mpe} yields for G_{ex}. Next, we argue why LJT^{mpe} is sound.

Theorem 2. *LJT^{mpe} is sound, i.e., computes an MPE for model G equivalent to an MPE computed for $gr(G)$.*

Proof. With a correct LJT, LJT^{mpe} constructs a valid jtree, the basis for local computations [16]. The *max-out* and *multiply* operators in the roles of marginalisation and combination fulfil the axioms for local computations in a probability

Algorithm 3. Outline of LVE^{mpe}	**Algorithm 4.** Outline of LJT^{mpe}
LVE^{MPE}(Model G, Evidence $\mathbf{E}$)	LJT^{MPE}(Model G, Evidence $\mathbf{E}$)
Absorb $\mathbf{E}$ in G	Construct FO jtree J
while G has not maxed out PRVs **do**	Enter $\mathbf{E}$ into J
if PRV A fulfils *max-out* prec. **then**	Pass max messages on J
Eliminate A using MAX-OUT	Get local model G_i from node i that received messages from all neighbours
else	▷ includes messages
Apply transformator	**return** LVE^{MPE}(G_i, ∅)
return MULTIPLY(G)	

propagation [16] that allow us to compute assignments locally and distribute them. Assuming that LVE^{mpe} is sound, computing assignments is sound. After one pass, the innermost node holds assignments for all model PRVs without evidence, which LJT returns.   □

3.3   Discussion

This section looks at MAP queries, a set of queries of different types as well as data and runtime performance.

Maximum A Posterior Queries: MAP, i.e., a query for a most probable assignment to a subset of model PRVs, is a more general case of MPE. The non-commutativity of summing out and maxing out leads to a restriction of the elimination order as it forces an algorithm to sum out randvars before maxing out query randvars. Logvars complicate matters further. Consider

$$\underset{(f,u)\in range(Fw(X,Y),User(X))}{\arg\max} \sum_{a\in range(Admin(Y))} \phi(u,a,f).$$

We need to sum out $Admin(Y)$ before maxing out $Fw(X,Y)$ and $User(X)$. As $Admin(Y)$ does not contain X and X is not count-convertible (it appears in two PRVs), we need to ground X to sum out $Admin(Y)$.

While MAP is harder to compute than MPE, FO jtrees allow for identifying harmless MAPs, namely, over whole parclusters. After message passing using LVE, a parcluster has all information for its PRVs with outside PRVs summed out. Using LVE^{mpe}, one can compute an MPE for a parcluster, producing an answer to a MAP over that parcluster. If a set of parclusters build a subtree in the FO jtree, LJT^{mpe} computes a MAP answer over the subtree.

Set of Queries: For a set of probability and assignment queries, LJT shows its particular strength. LVE computes an answer for each query starting with the original model and given evidence, using LVE^{mpe} for MPE queries. In contrast, LJT constructs an FO jtree for the input model and enters the given evidence. After message passing with LVE, all probability queries can be answered on smaller parcluster models. If an MPE query occurs, LJT does a message pass with LVE^{mpe}. For further probability queries, we either need to cache the old messages or repeat message passing with LVE. For MAP queries, LJT is able to use the LVE messages.

New evidence or a new model does not change the procedure for LVE as it starts with the full model and evidence. For LJT, new evidence means it needs to enter evidence anew and pass messages. A new model leads to constructing a new FO jtree followed by evidence entering and message passing.

Storage and Runtime: Regarding *storage*, switching from a probability to an assignment query requires storing assignments. This additional space requirement comes inherently with the new query type. Making the assignments part of the parfactors allows for dropping assignments no longer necessary during maxing out, setting space free as well. At the end, the MPE is directly encoded in the remaining parfactors to read out. Storing assignments as histograms allows for handling existential and universal quantification as one, reading out assignment counts directly from the histograms, and reducing constraints when logvars disappear through elimination. Of course, LJT requires more space for its FO jtree (independent of query types) trading off space with runtime.

Regarding *runtime*, MPE and probability queries for single randvars given evidence do not differ w.r.t. the magnitude of eliminations to perform. For MPEs, each remaining PRV has to be eliminated. For probability queries, each remaining PRV except the query randvar has to be eliminated. With just one MPE query, LJT^{mpe} incurs static overhead for constructing an FO jtree without trading it off further. Allowing a set of queries for MPEs, MAPs over parclusters, and probability distributions given a set of evidence, LJT trades off its overhead for construction and message passing easily. LJT re-uses the LVE messages to answer MAP queries, spending effort on one message pass for an MPE query.

In the worst case, MPE as well as probability queries have an exponential runtime even in the lifted case [18]. In such a case, LVE grounds all logvars, yielding computations identical to those performed in the propositional case.

4   Case Study: Risk Analysis

We present a case study on risk analysis of a network. The inherent uncertainties about attacks or causes within detecting if a network is compromised combined with dependencies and influences between different aspects of a network lend itself to model the scenario with a probabilistic model. A network also contains symmetries for various access points, hardware and software components, or users with different affiliations or permissions. Thus, parameterised models provide features to easily capture an attack scenario without an exploding number of randvars and thus, combinatorial effort during computations.

We expand our running example to model an AG close to [9] extending their case study with first-order constructs. In our scenario, logvars represent employees equipped with different permissions depending on their tasks (technician vs. accountant, admin vs. user). CRVs allow for modelling vulnerabilities where a certain number of components needs to be compromised. For the altered AG, Fig. 3 shows a corresponding FO jtree with seven nodes (without local models and separators). The grey node contains all servers in the network. The nodes with thick lines contain the three attacks.

Based on an AG, we can perform a risk analysis. A static analysis assess vulnerabilities in a network at rest. A dynamic analysis aims at identifying possible attack paths after an intrusion, which helps selecting countermeasures. Both analyses generate various queries, which makes LJT a suitable algorithm.

For a *static analysis*, one is interested in the probability of each network component to be compromised, which requires to compute probabilities for each PRV in the AG without evidence. To answer the queries, LJT passes messages on the FO jtree and answers queries for each PRV using a corresponding parcluster. Logvars allow for querying the likelihood of a certain number of users being compromised in the form of a conjunctive query. Changing employee numbers requires a new message pass but no new FO jtree.

Consider an intrusion detection system (IDS) detecting an intrusion. With the information from the IDS as evidence, a *dynamic analysis* generates various queries: (i) MPE to get the current most likely state of the system to find the nodes that are currently most likely compromised as well (undetected by the IDS), requiring one message pass; (ii) Probability of a specific attack being the source or the probability of specific nodes being compromised such as the firewall, requiring passing messages to answer the probability queries; (iii) MAP of, e.g., the servers parcluster provides the most likely state of the servers; (iv) Testing different attack paths by setting different nodes to compromised to track which nodes are now most likely to be compromised or by setting a suspected target to compromised to identify the nodes most likely compromised to reach the target, requiring passing messages for varying evidence sets. With only incremental changes to evidence, only one pass (instead of two) is necessary to propagate the changed evidence from its entry point to the remaining parts of the FO jtree.

We have implemented a prototype of LJT for probability and assignment queries using Taghipour's LVE implementation as a subroutine. LVE has faster runtimes with a single query. But with a second query, LJT has traded off its overhead and provides answers faster. A propositional version of the junction tree algorithm takes minutes with smaller domain sizes and runs into memory problems with the domain sizes given above. We use the AG in Fig. 3 with 20 parfactors and domain sizes $|\mathcal{D}(X)| = 90$, $|\mathcal{D}(X')| = 10$, $|\mathcal{D}(Y)| = 15$, and $|\mathcal{D}(Y')| = 3$ as input, yielding a grounded model size of 2599. LJT constructs the FO jtree in Fig. 3 in 16 ms.

We have tested the following queries

- MPE for the whole model
- MAP at parcluster $\mathbf{C}_5$: *Mailserver, Webserver, Dataserver*
- Probability query for *Attack1, Attack2, Mailserver, Webserver, Firewall, Dataserver,* $User(x_1)$, $User(x_1')$, $Admin(y_1)$, $Admin(y_1')$

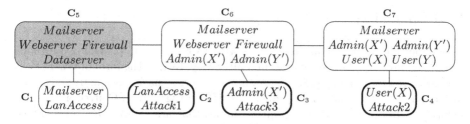

Fig. 3. FO jtree for an attack scenario (without local models and separators)

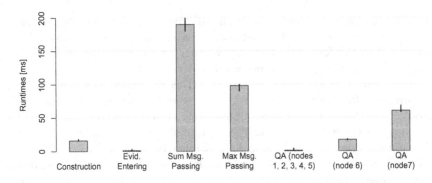

Fig. 4. Runtimes [ms] for the steps of LJT for the AG (whiskers min/max)

We have tested the following evidence

- $Attack2 = true$
- $DataServer = true$
- $User(X) = true$ for 30 instances of X
- $User(X') = true$ for 3 instances of X'

Figure 4 shows runtimes for the LJT steps averaged over several runs. Evidence entering with varying evidence does not take much time even if evidence affects each parcluster. Message passing takes the most time. For an MPE, this step is faster as only one pass is necessary. The time spent on messages is offset during QA. If increasing the domain sizes, the runtimes for inference rise linearly. The domain sizes do not affect the time for construction and evidence entering. Message passing and QA take up slightly more time. The time for answering a query varies between 0.1 and 60 ms, depending on the parclusters used for QA. The times for MAP versus a probability query do not differ on average.

If setting up new evidence and asking a query, LJT outputs an answer after 200–250 ms, which is in average the time LVE needs to answer a single query. After the first query, answers take between 0.1 and 60 ms again. So, even after setting up new evidence, runtimes are reasonably low for new queries.

5   Conclusion

This paper formalises computing MPEs with LVE and LJT, providing a formalism to compute assignment queries in a lifted way. LVE in the form of GC-FOVE and LJT are now able to compute probability queries (likelihood, marginals, conditionals) as well as assignment queries (MPE, MAP). An FO jtree allows for identifying harmless MAP assignment queries that are computable reusing the messages passed for probability queries. The case study on risk analysis in the area of IT security shows LVE and LJT in an applied context where different types of queries occur regularly. Using their formalism allows for capturing large scenarios while providing algorithms for fast query answering.

Future work includes adapting LJT to incrementally changing environments as to not restart FO jtree construction from scratch and salvage as much work as possible. Additionally, we look at ways to further optimise runtimes with parallelisation, caching, and indexing. Currently, we work on setting up a dynamic version that models the temporal or sequential aspect found in many applications.

References

1. Braun, T., Möller, R.: Lifted junction tree algorithm. In: Friedrich, G., Helmert, M., Wotawa, F. (eds.) KI 2016. LNCS (LNAI), vol. 9904, pp. 30–42. Springer, Cham (2016). https://doi.org/10.1007/978-3-319-46073-4_3
2. Ceylan, İ.İ., Borgwardt, S., Lukasiewicz, T.: Most probable explanations for probabilistic database queries. In: Proceedings of the 26th International Joint Conference on Artificial Intelligence (2017)
3. Chen, H., Erol, Y., Shen, E., Russell, S.: Probabilistic model-based approach for heart beat detection. Physiol. Measur. **37**(9), 1404 (2016)
4. Dawid, A.P.: Applications of a general propagation algorithm for probabilistic expert systems. Stat. Comput. **2**(1), 25–36 (1992)
5. Dechter, R.: Bucket elimination: a unifying framework for probabilistic inference. In: Learning and Inference in Graphical Models, pp. 75–104. MIT Press (1999)
6. Gribkoff, E., van den Broeck, G., Suciu, D.: The most probable database problem. In: Proceedings of the 1st International Workshop on Big Uncertain Data (2014)
7. Lauritzen, S.L., Spiegelhalter, D.J.: Local computations with probabilities on graphical structures and their application to expert systems. J. R. Stat. Soc. Ser. B: Methodol. **50**, 157–224 (1988)
8. Milch, B., Zettelmoyer, L.S., Kersting, K., Haimes, M., Kaelbling, L.P.: Lifted probabilistic inference with counting formulas. In: Proceedings of the 23rd Conference on Artificial Intelligence, AAAI 2008, pp. 1062–1068 (2008)
9. Muñoz-González, L., Sgandurra, D., Barrère, M., Lupu, E.C.: Exact inference techniques for the analysis of Bayesian attack graphs. IEEE Trans. Dependable Secure Comput. **PP**(99), 1–14 (2017)
10. Nilsson, D.: An efficient algorithm for finding the M most probable configurations in probabilistic expert systems. Stat. Comput. **8**(2), 159–173 (1998)
11. Pearl, J.: Probabilistic Reasoning in Intelligent Systems: Networks of Plausible Inference. Morgan Kaufmann, Burlington (1988)
12. Poole, D.: First-order probabilistic inference. In: Proceedings of the 18th International Joint Conference on Artificial Intelligence, IJCAI 2003 (2003)
13. de Salvo Braz, R.: Lifted first-order probabilistic inference. Ph.D. thesis, University of Illinois at Urbana Champaign (2007)
14. de Salvo Braz, R., Amir, E., Roth, D.: MPE and partial inversion in lifted probabilistic variable elimination. In: Proceedings of the 21st Conference on Artificial Intelligence, AAAI 2006 (2006)
15. Schröder, M., Lüdtke, S., Bader, S., Krüger, F., Kirste, T.: LiMa: sequential lifted marginal filtering on multiset state descriptions. In: Kern-Isberner, G., Fürnkranz, J., Thimm, M. (eds.) KI 2017. LNCS (LNAI), vol. 10505, pp. 222–235. Springer, Cham (2017). https://doi.org/10.1007/978-3-319-67190-1_17
16. Shenoy, P.P., Shafer, G.R.: Axioms for probability and belief-function propagation. Uncertain. Artif. Intell. **4**(9), 169–198 (1990)

17. Shterionov, D., Renkens, J., Vlasselaer, J., Kimmig, A., Meert, W., Janssens, G.: The most probable explanation for probabilistic logic programs with annotated disjunctions. In: Davis, J., Ramon, J. (eds.) ILP 2014. LNCS (LNAI), vol. 9046, pp. 139–153. Springer, Cham (2015). https://doi.org/10.1007/978-3-319-23708-4_10

18. Taghipour, N., Davis, J., Blockeel, H.: First-order decomposition trees. In: Advances in Neural Information Processing Systems 26, pp. 1052–1060. Curran Associates, Inc. (2013)

19. Taghipour, N., Fierens, D., Davis, J., Blockeel, H.: Lifted variable elimination: decoupling the operators from the constraint language. J. Artif. Intell. Res. **47**(1), 393–439 (2013)

20. Zhang, N.L., Poole, D.: A simple approach to Bayesian network computations. In: Proceedings of the 10th Canadian Conference on Artificial Intelligence, pp. 171–178 (1994)

Lifted Dynamic Junction Tree Algorithm

Marcel Gehrke$^{(\boxtimes)}$, Tanya Braun, and Ralf Möller

Institute of Information Systems, Universität zu Lübeck, Lübeck, Germany
{gehrke,braun,moeller}@ifis.uni-luebeck.de

Abstract. Probabilistic models involving relational and temporal aspects need exact and efficient inference algorithms. Existing approaches are approximative, include unnecessary grounding, or do not consider the relational and temporal aspects of the models. One approach for efficient reasoning on relational static models given multiple queries is the lifted junction tree algorithm. In addition, for propositional temporal models, the interface algorithm allows for efficient reasoning. To leverage the advantages of the two algorithms for relational temporal models, we present the lifted dynamic junction tree algorithm, an exact algorithm to answer multiple queries efficiently for probabilistic relational temporal models with known domains by reusing computations for multiple queries and multiple time steps. First experiments show computational savings while appropriately accounting for relational and temporal aspects of models.

1   Introduction

Areas like healthcare and logistics involve probabilistic data with relational and temporal aspects and need efficient exact inference algorithms, e.g., as indicated by Vlasselaer et al. [23]. These areas involve many objects in relation to each other with changing information over time and uncertainties about objects, objects attributes, or relations. More specifically, healthcare systems involve electronic health records (the relational part) for many patients (the objects), streams of measurements over time (the temporal part), and uncertainties due to, e.g., missing or incomplete information, for example caused by data integration of records from different hospitals. By performing model counting, probabilistic databases (PDBs) can answer huge queries, which embed most of the model behaviour, for relational temporal models with uncertainties [5,6]. However, we build more expressive and compact models including behaviour (offline) enabling efficient answering of smaller queries (online). To be more precise, for query answering we perform deductive reasoning by computing marginal distributions at discrete time steps. In this paper we study the problem of exact inference, in form of filtering and prediction, in large temporal models that exhibit symmetries.

This research originated from the Big Data project being part of Joint Lab 1, funded by Cisco Systems Germany, at the centre COPICOH, University of Lübeck.

© Springer International Publishing AG, part of Springer Nature 2018
P. Chapman et al. (Eds.): ICCS 2018, LNAI 10872, pp. 55–69, 2018.
https://doi.org/10.1007/978-3-319-91379-7_5

For exact inference on propositional temporal models, a naive approach is to unroll the temporal model for a given number of time steps and use any exact inference algorithm for static, i.e., non-temporal, models. In the worst case, once the number of time steps changes, one has to unroll the model and infer again. To prevent the complete unrolling, Murphy proposes the interface algorithm [13].

One reasoning approach for leveraging the relational aspect of a static model is first-order probabilistic inference. For models with known domain size, it exploits symmetries in a model by combining instances to reason with representatives, known as lifting [16]. Poole introduces parametric factor graphs as relational models and proposes lifted variable elimination (LVE) as an exact inference algorithm on relational models [16]. Further, de Salvo Braz [18], Milch et al. [11], and Taghipour et al. [20] extend LVE into its current form. Lauritzen and Spiegelhalter [9] introduce the junction tree algorithm. To benefit from the ideas of the junction tree algorithm and LVE, Braun and Möller [2] present the lifted junction tree algorithm (LJT) that efficiently performs exact first-order probabilistic inference on relational models given a set of queries.

We aim at an exact and efficient inference algorithm for a set of queries that handles both the relational and the temporal aspect. To this end, we present the lifted dynamic junction tree algorithm (LDJT) that combines the advantages of the interface algorithm and LJT. Specifically, this paper contributes (i) a definition for parameterised probabilistic dynamic models (PDMs) as a representation for relational temporal models, and (ii) a formal description of LDJT, a reasoning algorithm for PDMs, a set of queries, and a set of observations (evidence).

Related work for inference on relational temporal models mostly consists of approximative approaches. Additionally, to being approximative, these approaches involve unnecessary groundings or are only designed to handle single queries efficiently. Ahmadi et al. propose a lifted (loopy) belief propagation [1]. From a factor graph, they build a compressed factor graph and apply lifted belief propagation with the idea of the factored frontier algorithm [12], which is an approximate counterpart to the interface algorithm [13]. Thon et al. introduce CPT-L, a probabilistic model for sequences of relational state descriptions with a partially lifted inference algorithm [21]. Geier and Biundo present an online interface algorithm for dynamic Markov logic networks (DMLNs) [7], similar to the work of Papai et al. [15]. Both approaches slice DMLNs to run well-studied static MLN [17] inference algorithms on each slice individually. Further, the interface algorithm also slices the model to utilise static approaches. Two ways of performing online inference using particle filtering are described in [10,14].

Vlasselaer et al. introduce an exact approach for relational temporal models involving computing probabilities of each possible interface assignment [22].

LDJT has several benefits: The lifting approach exploits symmetries in the model to reduce the number of instances to perform inference on. For answering multiple queries, the junction tree idea enhances efficiency by clustering a model into submodels sufficient for answering a particular query. Further, the interface idea drastically reduces the size of the model and allows adding time steps dynamically, by an efficient separation of time steps. Furthermore, the junction tree structure of the model is reused for all time steps $t > 0$.

The remainder of this paper has the following structure: We begin by introducing PDMs to represent relational temporal models. Followed by, LDJT an efficient reasoning algorithm for PDMs and evaluate LDJT compared to LJT and a ground interface approach. We conclude by looking at possible extensions.

2 Parameterised Probabilistic Dynamic Models

We introduce parameterised probabilistic models (PMs), which is mainly based on [3], as a representation for relational static models. Afterwards, we extend PMs to the temporal case, resulting in PDMs for relational temporal models.

2.1 Parameterised Probabilistic Models

PMs combine first-order logic with probabilistic models, representing first-order constructs using logical variables (logvars) as parameters.

Definition 1. *We define a basic block with $\mathbf{L}$ as a set of logvar names, Φ as set of factor names, and $\mathbf{R}$ a set of random variable (randvar) names. A parameterised randvar (PRV) $A = P(X^1, ..., X^n)$ represents a set of randvars behaving identically by combining a randvars $P \in \mathbf{R}$ with $X^1, ..., X^n \in \mathbf{L}$. If $n = 0$, the PRV is parameterless. The domain of a logvar L is denoted by $\mathcal{D}(L)$. The term range(A) provides possible values of a PRV A. Constraint $(\mathbf{X}, C_{\mathbf{X}})$ allows to restrict logvars to certain domain values and is a tuple with a sequence of logvars $\mathbf{X} = (X^1, ..., X^n)$ and a set $C_{\mathbf{X}} \subseteq \times_{i=1}^{n} \mathcal{D}(X^i)$. The symbol $\top$ denotes that no restrictions apply and may be omitted. The term $lv(Y)$ refers to the logvars in some element Y. The term $rv(Y)$ refers to the logvars in Y. The term $gr(Y)$ denotes the set of instances of Y with all logvars in Y grounded w.r.t. constraints.*

Now, we illustrate PMs with an example, which has the goal to remotely infer the condition of patients with regards to water retaining. To determine the condition of patients, we use the change of their weights. Further, we are interested in the condition of people the patient is living with, giving us indications to improve the inferred conditions. In case patients are living together and both are gaining weight, they probably overeat and do not retain water. If no new weights are submitted, we are interested whether the scale broke or the patient stopped submitting weights. In case only one patient stops to send weights, it is likely that the patient stopped deliberately. If patients living together stop to submit weights at the same time, it is more likely that their scale broke. Hence, we can improve the accuracy of inference by accounting for patients living together.

To model the example, we use the randvar names C, LT, S, and W for Condition, LivingTogether, ScaleWorks, and Weight, respectively, and the logvar names X and X'. From the names, we build PRVs $C(X)$, $LT(X, X')$, $S(X)$, and $W(X)$. The domain of X and X' is the set $\{alice, bob, eve\}$. The range of $C(X)$ is $\{normal, deviation, retains\ water, stopped\}$, $LT(X, X')$ and $S(X)$ have range {*true, false*}, and of $W(X)$ has range {*steady, falling, rising, null*}. Now, we define parametric factors (parfactors), to set PRVs into relation to each other.

Definition 2. *We define a parfactor g with $\forall \mathbf{X} : \phi(\mathcal{A}) \,|C.$ $\mathbf{X} \subseteq \mathbf{L}$ being a set of logvars over which the factor generalises and $\mathcal{A} = (A^1, ..., A^n)$ a sequence of PRVs. We omit $(\forall \mathbf{X} :)$ if $\mathbf{X} = lv(\mathcal{A})$. A function $\phi : \times_{i=1}^{n} range(A^i) \mapsto \mathbb{R}^+$ with name $\phi \in \Phi$ is identically defined for all grounded instances of $\mathcal{A}$. A list of all input-output values is the complete specification for ϕ. The output value is called potential. C is a constraint on $\mathbf{X}$. A PM (model) $G := \{g^i\}_{i=0}^{n-1}$ is a set of parfactors and represents the full joint probability distribution $P(G) = \frac{1}{Z} \prod_{f \in gr(G)} \phi(\mathcal{A}_f)$ where Z is a normalisation constant.*

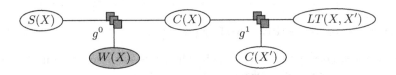

Fig. 1. Parfactor graph for model G^{ex} with observable nodes in grey

Now, we build the model G^{ex} of our example with the parfactors:

$$g^0 = \phi^0(C(X), S(X), W(X))|\top \text{ and } g^1 = \phi^1(C(X), C(X'), LT(X, X'))|\kappa^1$$

We omit the concrete mappings of ϕ^0 and ϕ^1. Parfactor g^0 has the constraint $\top$, meaning it holds for *alice, bob,* and *eve.* The constraint κ^1 of g^1 ensures that $X \neq X'$ holds. Figure 1 depicts G^{ex} as a parfactor graph and shows PRVs as nodes, which are connected via undirected edges to nodes of parfactors in which they appear. We can observe the weight of patients. The remaining PRVs are latent.

The semantics of a model is given by grounding and building a full joint distribution. In general, queries ask for a probability distribution of a randvar using a model's full joint distribution and given fixed events as evidence.

Definition 3. *Given a PM G, a ground PRV Q and grounded PRVs with fixed range values $\mathbf{E}$ the expression $P(Q|\mathbf{E})$ denotes a query w.r.t. $P(G)$.*

In our example, a query is $P(C(bob)|W(bob) = steady)$, asking for the probability distribution of *bob*'s condition given information about his weight.

2.2 Parameterised Probabilistic Dynamic Models

To define PDMs, we use PMs and the idea of how Bayesian networks (BNs) give rise to dynamic Bayesian networks (DBNs). We define PDMs based on the first-order Markov assumption, i.e., a time slice t only depends on the previous time slice $t - 1$. Further, the underlining process is stationary, i.e., the model's behaviour does not change over time.

Definition 4. *A PDM is a pair of PMs* $(G_0, G_\rightarrow)$ *where* G_0 *is a PM representing the first time step and* $G_\rightarrow$ *is a two-slice temporal parameterised model (2TPM) representing* $\mathbf{A}_{t-1}$ *and* $\mathbf{A}_t$ *with* $\mathbf{A}_t$ *a set of PRVs from time slice t.*

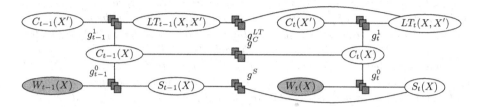

Fig. 2. $G_\rightarrow^{ex}$ the two-slice temporal parfactor graph for model G^{ex}

Figure 2 shows how the model G^{ex} behaves over time. $G_\rightarrow^{ex}$ consists of G^{ex} for time step $t - 1$ and for time step t with inter-slice parfactors for the behaviour over time. In this example, the latent PRVs depend on each other from one time slice to the next, which is represented with the parfactors g^{LT}, g^C, and g^S.

Definition 5. *Given a PDM G, a ground PRV* Q_t *and grounded PRVs with fixed range values* $\mathbf{E}_{\{0:t\}}$ *the expression* $P(Q_t | \mathbf{E}_{\{0:t\}})$ *denotes a query w.r.t.* $P(G)$.

The problem of answering queries for the current time step is called filtering and for a time step in the future it is called prediction.

3 Lifted Dynamic Junction Tree Algorithm

We start by recapping LJT to provide means to answer queries for PMs, mainly based on [3], and the interface algorithm, an approach to perform inference for propositional temporal models, mainly based on [13]. Afterwards, we present LDJT, consisting of a first-order junction tree (FO jtree) construction for a PDM and an efficient reasoning algorithm to perform filtering and prediction.

3.1 Lifted Junction Tree Algorithm

LJT provides efficient means to answer a set of queries $\{P(Q^i | \mathbf{E})\}_{i=1}^k$ given a PM G and evidence $\mathbf{E}$, by performing the following steps: (i) Construct an FO jtree J for G. (ii) Enter $\mathbf{E}$ in J. (iii) Pass messages. (iv) Compute answer for each query Q^i. We first define an FO jtree and then go through each step. For an FO jtree, we need parameterised clusters (parclusters), the nodes of an FO jtree.

Definition 6. *A parcluster* **C** *is defined by* $\forall \mathbf{L} : \mathbf{A}|C$. **L** *is a set of logvars,* **A** *is a set of PRVs with* $lv(\mathbf{A}) \subseteq \mathbf{L}$, *and* C *a constraint on* **L**. *We omit* $(\forall \mathbf{L} :)$ *if* $\mathbf{L} = lv(\mathbf{A})$. *A parcluster* $\mathbf{C}^i$ *can have parfactors* $\phi(\mathcal{A}^\phi)|C^\phi$ *assigned given that (i)* $\mathcal{A}^\phi \subseteq \mathbf{A}$, *(ii)* $lv(\mathcal{A}^\phi) \subseteq \mathbf{L}$, *and (iii)* $C^\phi \subseteq C$ *hold. We call the set of assigned parfactors a local model* G^i.

An FO jtree for a model G *is* $J = (\mathbf{V}, \mathbf{E})$ *where* J *is a cycle-free graph, the nodes* **V** *denote a set of parcluster, and the set of edges* **E** *the paths between parclusters. An FO jtree must satisfy the following three properties: (i) A parcluster* $\mathbf{C}^i$ *is a set of PRVs from* G. *(ii) For each parfactor* $\phi(\mathcal{A})|C$ *in* G, $\mathcal{A}$ *must appear in some parcluster* $\mathbf{C}^i$. *(iii) If a PRV from* G *appears in two parclusters* $\mathbf{C}^i$ *and* $\mathbf{C}^j$, *it must also appear in every parcluster* $\mathbf{C}^k$ *on the path connecting nodes* i *and* j *in* J. *The separator* $\mathbf{S}^{ij}$ *containing shared PRVs of edge* $i - j$ *in* J *is given by* $\mathbf{C}^i \cap \mathbf{C}^j$. *The FO jtree. is minimal if by removing a PRV from any parcluster, the FO jtree stops being an FO jtree.*

LJT constructs an FO jtree using a first-order decomposition tree (FO dtree). Analogous to the ground case [4], LJT can use the clusters of an FO dtree to construct an FO jtree. For the details on construction of an FO dtree from a PM, the formal definition, and properties of an FO dtree, refer to Taghipour et al. [19]. For our approach, important FO dtree characteristics are: (i) each leaf node contains exactly one parfactor, (ii) the cluster for a leaf node l consists of the randvars of the corresponding parfactor in l, $rv(l)$, and (iii) a (par)cluster in an (FO) jtree corresponds to a cluster of an (FO) dtree.

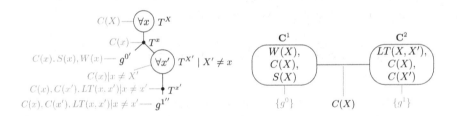

Fig. 3. FO dtree for model G^{ex} **Fig. 4.** Minimised FO jtree for model G^{ex}

Figure 3 shows the FO dtree for the model G^{ex}, with the clusters of the FO dtree for every node in grey. An FO jtree directly constructed from clusters of an FO dtree is non-minimal. To minimise an FO jtree, LJT merges neighbouring parclusters if one parcluster is the subset of the other. The parfactors at the leaf nodes from an FO dtree determine the local models for parclusters.

LJT enters evidence in the FO jtree and passes messages through an *inbound* and an *outbound* pass, to distribute local information of the nodes through the FO jtree. To compute a message, LJT eliminates all non-seperator PRVs from the parcluster's local model and received messages. After message passing, LJT answers queries. For each query, LJT finds a parcluster containing the query term and sums out all non-query terms in its local model and received messages.

Figure 4 shows the minimised FO jtree corresponding to the FO dtree from Fig. 3. One possibility to obtain the FO jtree is to merge the clusters of T^X into T^x and then T^x into the leaf with $g^{0'}$ and the remaining clusters of the FO dtree into the leaf with $g^{1''}$. By merging the clusters of the FO dtree, LJT acquires a minimised FO jtree. Here, each parfactor from the PM makes up the local model of a parcluster, the ideal case to answer queries. Throughout the paper, we also have FO jtrees with more parfactors in a local model of a parcluster, resulting in less messages during message passing but higher query answering efforts.

Additionally, Fig. 4 shows the local models of the parclusters and the separator PRV $C(X)$ as label of the edge. Thus, we have two parclusters, $\mathbf{C}^1$ and $\mathbf{C}^2$, in the minimised FO jtree. Before LJT answers queries, it passes messages to account for evidence. During the *inbound* phase LJT sends messages from $\mathbf{C}^1$ to $\mathbf{C}^2$ and from $\mathbf{C}^2$ to $\mathbf{C}^1$. If we want to know whether the scale from *alice* works, we have to query for $P(S(alice))$ for which LJT can use parcluster $\mathbf{C}^1$. LJT sums out $C(X)$, $W(X)$, and $S(X)$ where $X \neq alice$ from $\mathbf{C}^1$'s local model G^{ex^1} combined with the received messages, here, the one message from $\mathbf{C}^2$.

3.2 Interface Algorithm

The interface algorithm for DBNs allows to efficiently pass on the current state of the model from one time slice to the next, while being able to make use of a static junction tree algorithm for BNs. The interface algorithm defines the set of nodes with outgoing edges to the next time slice as an interface for temporal d-separation. The interface has to be in one cluster of the associated jtree. While proceeding to the next time step, the interface algorithm reuses the structure of the jtree and only passes in information from the outgoing interface cluster.

A DBN is defined using two BNs: B_0 is a BN which defines the prior and $B_\rightarrow$ is a two-slice temporal Bayesian network (2TBN), which models the temporal behaviour. The interface algorithm uses that the set of nodes with outgoing edges, I_t, to the next time slice from $B_\rightarrow$ is sufficient to d-separate the past from the future. The interface algorithm builds a jtree J_0 for B_0 and ensures during the creation that I_0 ends up in a cluster of the jtree. The cluster containing I_0 is labeled *in-* and *out-cluster*. Then, the algorithm turns $B_\rightarrow$ into a 1.5TBN, H_t (H for half), by removing all non interface nodes N_{t-1} and their edges from the first slice of $B_\rightarrow$. Now, it constructs a jtree J_t for H_t and ensures that I_{t-1} and I_t each end up in clusters of the jtree. The cluster containing I_{t-1} is labeled *in-cluster* and the cluster containing I_t is labeled *out-cluster*. The idea is to pass messages α_{t-1} over I_{t-1} from the *out-cluster* of J_{t-1} to the *in-cluster* of J_t.

Figure 5 shows how the interface algorithm uses the *in-* and *out-clusters* of the jtrees J_0 and J_t for the first three time steps to pass on the current state in each time step. To reason for $t = 0$, the interface algorithm uses J_0. First, a junction tree algorithm enters evidence in J_0 if available, passes messages, and answers queries. The interface algorithm then computes a message using the *out-cluster* of J_0, by summing out all non-interface variables, to pass the message on via the separator to J_1. For all $t > 0$, the interface algorithm instantiates J_t for that time step. Afterwards, the interface algorithm recovers the state of the model by

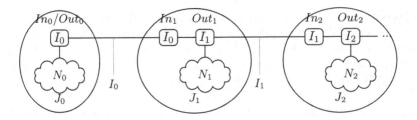

Fig. 5. Combination of jtrees using interface algorithm

adding the message from the *out-cluster* of J_{t-1} to the *in-cluster* of J_t. For $t = 1$ the interface algorithm instantiates J_1 and then adds the message containing I_0 to J_1's *in-cluster*. After the interface algorithm recovered the previous state by adding the message, it behaves as it did for $t = 0$. A junction tree algorithm enters evidence in J_1 if available, passes messages, and answers queries. During message passing, information from I_0 is distributed through the jtree J_1 and hence present in the *out-cluster* to compute the message over I_1.

3.3 LDJT: Overview

LDJT efficiently answers sets of queries $\{P(Q_t^i|\mathbf{E}_t)\}_{i=1}^k$, $Q_t^i \in \{\mathbf{Q}_t\}_{t=0}^T$, given a PDM G and evidence $\{\mathbf{E}_t\}_{t=0}^T$, by performing the following steps:

(i) Offline construction of the two FO jtrees J_0 and J_t with *in-* and *out-clusters*
(ii) For $t = 0$, using J_0 to enter $\mathbf{E}_0$, pass messages, answer $\mathbf{Q}_0$, and preserve the state in message α_0
(iii) For $t > 0$ instantiate J_t, recover the previous state from message α_{t-1}, enter $\mathbf{E}_t$ in J_t, pass messages, answer $\mathbf{Q}_t$, and preserve the state in message α_t

LDJT solves the filtering and prediction problems efficiently by reusing a compact structure for multiple queries and time steps. Further, LDJT only requires the current evidence and the state of the interface from the previous time step for queries. Next, we show how LDJT constructs the FO jtrees J_0 and J_t with *in-* and *out-clusters* and then, how LDJT uses the FO jtrees for reasoning.

3.4 LDJT: FO Jtree Construction for PDMs

The steps of FO jtree constructions are shown in Algorithm 1. LDJT constructs an FO jtree for G_0 and for $G_\rightarrow$, both with an incoming and outgoing interface. Therefore, LDJT first identifies the interface PRVs $\mathbf{I}_t$ for a time slice t. We define $\mathbf{I}_{t-1}$ as follows:

Definition 7. *The forward interface is defined as* $\mathbf{I}_{t-1} = \{A_{t-1}^i \mid \exists\phi(\mathcal{A})|C \in G : A_{t-1}^i \in \mathcal{A} \land \exists A_t^j \in \mathcal{A}\}$, *i.e., the PRVs which have successors in the next slice. The set of non-interface PRVs is* $\mathbf{N}_t = \mathbf{A}_t \setminus \mathbf{I}_t$.

To ensure interface PRVs **I** ending up in a single parcluster, LDJT adds a parfactor g^I over the interface to the model with uniform potentials in the mappings, e.g., mapping all input values to 1. $\mathbf{I}_0$ has to be assigned to a local model of a parcluster of J_0. To ensure that $\mathbf{I}_0$ ends up in a single parcluster, LDJT adds a parfactor g_0^I over $\mathbf{I}_0$ to G_0. When LDJT constructs the FO jtree, it first constructs an FO dtree. In the FO dtree, the parfactor ends up as a leaf and the cluster of each leaf contains the randvars of the corresponding parfactor. Further, while minimising the FO jtree, the parfactor is assigned to a local model of a parcluster. Thereby, by adding the interface parfactor to the model ensures that the resulting FO jtree has an incoming and outgoing interface. LDJT labels the parcluster with g_0^I from J_0 as *in-* and *out-cluster*.

For $G_{\rightarrow}^{ex}$, which is shown in Fig. 2, PRVs $C_{t-1}(X)$, $S_{t-1}(X)$, and $LT_{t-1}(X, X')$ have children in the next time slice, making up $\mathbf{I}_{t-1}$. Figure 6 shows G_0^{ex} with g_0^I, by setting $t-1 = 0$ and removing all other PRVs and parfactors that do not belong to $t-1$, leaving us with the PRVs, parfactors, and edges with thicker lines. Figure 7 shows a corresponding FO jtree, with *in-* and *out-cluster* labelling. In the FO jtree, parcluster $\mathbf{C}_0^2$ is the only candidate to be the *in-* and *out-cluster*.

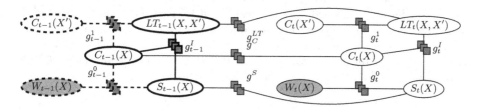

Fig. 6. Two-slice temporal parfactor graph for model $G_{\rightarrow}^{ex}$ with interface parfactors

Having J_0, LDJT constructs J_t for the remaining time steps. During inference, time slice $t-1$ will encode the past and the actual inference is performed on t, allowing us to transform the 2TPM into a 1.5-slice TPM H_t.

Algorithm 1. FO Jtree Construction for a PDM $(G_0, G_{\rightarrow})$

---

  function DFO-JTREE($G_0, G_{\rightarrow}$)
    $\mathbf{I}_t$:= Set of interface PRVs for time slice t
    g_0^I := Parfactor for $\mathbf{I}_0$
    G_0 := $g_0^I \cup G_0$
    J_0 := Construct minimized FO jtree for G_0
    g_{t-1}^I := Parfactor for $\mathbf{I}_{t-1}$
    g_t^I := Parfactor for $\mathbf{I}_t$
    H_t := $\{\phi(\mathcal{A}) | C \in G_{\rightarrow} \mid \forall A \in \mathcal{A} : A \notin \mathbf{N}_{t-1}\}$
    G_t := $(g_{t-1}^I \cup g_t^I \cup H_t)$
    J_t := Construct minimized FO jtree for G_t
  return $(J_0, J_t, \mathbf{I}_t)$

Definition 8. $H_t = \{\phi(\mathcal{A})|C \in G_\rightarrow \mid \forall A \in \mathcal{A} : A \notin \mathbf{N}_{t-1}\}$, *i.e., eliminating all non-interface PRVs, their parfactors, and edges from the first time slice in* $G_\rightarrow$.

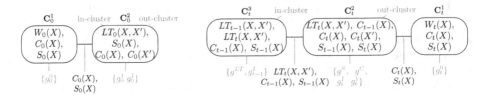

Fig. 7. FO jtree J_0 for G_0^{ex} Fig. 8. FO jtree J_t for $G_\rightarrow^{ex}$

LDJT needs to ensure that in the resulting FO jtree from $G_\rightarrow$, $\mathbf{I}_{t-1}$ and $\mathbf{I}_t$ each end up in parclusters. Hence, LDJT adds parfactors g_{t-1}^I and g_t^I to $G_\rightarrow$. LDJT constructs J_t from H_t with g_{t-1}^I and g_t^I and labels the parcluster containing g_{t-1}^I as *in-cluster* and the parcluster containing g_t^I as *out-cluster*.

Figure 6 shows $G_\rightarrow^{ex}$ with g_{t-1}^I and g_t^I added. By removing all nodes, parfactors, and edges with dashed lines the result is a 1.5-slice TPM of $G_\rightarrow^{ex}$ with g_{t-1}^I and g_t^I added. Figure 8 shows a corresponding FO jtree. To label the *in-cluster*, LDJT searches for a parcluster with g_{t-1}^I in its local model, which $\mathbf{C}_t^3$ has, and labels the parcluster as *in-cluster*. LDJT does the same for g_t^I and labels $\mathbf{C}_t^2$ as *out-cluster*.

3.5   LDJT: Reasoning with FO Jtrees from PDMs

Algorithm 2 provides the steps for answering queries. Since J_0 and J_t are static, LDJT uses LJT as a subroutine passing on an already constructed FO jtree, queries, and evidence for step t and lets LJT handle evidence entering, message passing, and query answering. For the first time step $t = 0$, LDJT uses J_0, takes the queries and evidence for $t = 0$ and uses LJT to answer the queries. After all queries are answered, LDJT needs to preserve the current state to pass it on to the next time slice. Therefore, LDJT uses the *out-cluster* parcluster, sums out all non-interface PRVs from that parcluster, and saves the result in message α_0, which holds the state of the PRVs that have an impact on the next time slice and thus, encodes the current state. Afterwards, LDJT increases t by one.

For time steps $t > 0$, LDJT uses J_t and first recovers the state of the previous time step by adding α_{t-1} to the *in-cluster* of J_t. LDJT then uses LJT to perform filtering. During message passing of LJT, information from the previous state is distributed through the FO jtree. After query answering, LDJT sums out all non-interface PRVs from the *out-cluster* of J_t, saves the result in message α_t, and increases t. Using the interface clusters, the FO jtrees are m-separated from one time step to the next and LDJT can use J_t once constructed for all $t > 0$.

Figure 9 depicts how LDJT uses the interface to pass on the current state from time step three to four. First, LDJT enters evidence for $t = 3$ using LJT.

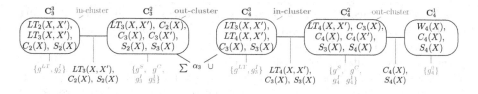

Fig. 9. LDJT passes on the current state to the next time step, J_3 shown without $\mathbf{C}_3^1$

Afterwards, LJT distributes local information by message passing. To capture the state at $t = 3$, LDJT needs to sum out the non-interface PRVs $C_2(X)$, $C_3(X')$, and $S_2(X)$ from $\mathbf{C}_3^2$ and save the result in message α_3. Thus, LDJT sums out the non-interface PRVs of the parfactors g^S, g^C, g_3^I, g_3^1 and the received messages m_3^{12} and m_3^{32}. After increasing t by one, LDJT adds α_3 to the *in-cluster* of J_4, $\mathbf{C}_4^3$. α_3 is then distributed by message passing and accounted for in α_4.

Given a stream of evidence, a stream of queries, which can be the same for each time step, and a PDM, LDJT performs filtering and prediction. To perform filtering, LDJT enters the current evidence, e.g., the patients weight, into the current FO jtree, which already accounts for the past, and answers queries, e.g. what is the condition of a patient. Further, LDJT performs prediction, for example given the evidence so far what is the condition of the patient in 10 time steps. To perform prediction, LDJT has to enter the current evidence, e.g. the patients weight, in the current FO jtree and passes on this information through all FO jtrees until LDJT reaches the time step the query is designated for and then answers the queries. LDJT efficiently solves the problem of performing filtering and prediction for PDMs by using a compact structure, which is also reused for all $t > 0$. Further, LDJT only calculates one additional message per time step for temporal m-separation.

Theorem 1. *LDJT is correct regarding filtering and prediction*

Algorithm 2. LDJT Alg. for PDM $(G_0, G_\to)$, Queries $\{\mathbf{Q}\}_{t=0}^T$, Evidence $\{\mathbf{E}\}_{t=0}^T$

---

procedure LDJT$(G_0, G_\to, \{\mathbf{Q}\}_{t=0}^T, \{\mathbf{E}\}_{t=0}^T)$

   $t := 0$

   $(J_0, J_t, \mathbf{I}_t) := \text{DFO-JTREE}(G_0, G_\to)$

   while $t \neq T$ **do**

      if $t = 0$ **then**

         $J_0 := \text{LJT}(J_0, \mathbf{Q}_0, \mathbf{E}_0)$ ▷ Including query answering, no FO jtree const.

         $\alpha_0 := \sum_{J_0(\text{out-cluster})\backslash\mathbf{I}_0} J_0(\text{out-cluster})$

         $t := t + 1$

      else

         $J_t(\text{in-cluster}) := \alpha_{t-1} \cup J_t(\text{in-cluster})$

         $J_t := \text{LJT}(J_t, \mathbf{Q}_t, \mathbf{E}_t)$ ▷ Including query answering, no FO jtree const.

         $\alpha_t := \sum_{J_t(\text{out-cluster})\backslash\mathbf{I}_t} J_t(\text{out-cluster})$

         $t := t + 1$

Proof. The interface PRVs m-separate the time slices for a given PDM. Using interface parfactors, LDJT ensures that the FO jtrees for the initial time step and the copy pattern $G_\rightarrow$ have an *in-cluster* and an *out-cluster*. The interface parfactors have uniform potentials in the mappings, therefore, have no impact on message passing or calculating answers to queries, besides a scaling factor. Further, the interface message α_t is equivalent to having the PDM unrolled for t time steps with evidence entered for each time step and calculating a query over the interface. To perform filtering for $t+1$, LDJT uses LJT to distribute the information contained in α_t, which accounts for all evidence until time step t, and the entered evidence for time step $t+1$ in J_{t+1} during the *inbound* and *outbound* phase of message passing. Hence, all parclusters of J_{t+1} receive information accounting for all evidence until time step $t+1$. Therefore, LDJT can use J_{t+1} to perform filtering for $t+1$ and prediction can be reformulated as filtering without new evidence added.

4   Evaluation

For different maximum time steps and domain sizes, we evaluate the largest (par)cluster and the number of (par)clusters in the (FO) jtrees for LDJT, LJT based on an unrolled model, and a ground interface approach, named JT. The number of (par)clusters n in an (FO) jtree determines the number of messages calculated during message passing. In general, message passing consists of calculating $2 \cdot (n-1)$ messages. Given new evidence, e.g., for a new time step, we need to calculate and parse new messages. The number of (P)RVs m in the largest (par)cluster, indicates how many variables we need to sum out, namely in the worst case we need to sum out $m-1$ (P)RVs for each message. Further, the largest (par)cluster and the number of (par)clusters can also be used to determine the complexity of VE and LVE a priori.

For G^{ex} Fig. 10 depicts the number of (par)clusters and Fig. 11 shows the number of (P)RVs in the largest (par)cluster for LDJT, LJT, and JT, with a domain size $\mathcal{D}(X) = \mathcal{D}(X') = 32$, for different maximum time steps. In these figures, we see that with increasing maximum time steps, the size of the FO jtree and the number of PRVs in the largest parcluster increase for LJT, while they remain constant for LDJT and JT. With increasing time steps, the unrolled model becomes larger. Therefore, the size input model increases with the time steps for LJT, while the input model remains constant for LDJT and JT.

For different domain sizes, Fig. 12 shows the number of (par)clusters and Fig. 13 depicts the number of (P)RVs in the largest (par)cluster for LDJT, LJT for 32 time steps, and JT. In these figures, we can see that with increasing domain sizes, the size of the jtree and the number of RVs in the largest cluster increase for JT, while they remain constant for LDJT and LJT. Due to more groundings, the input model increases with the domain size for JT. LDJT and LJT combine the instances and handle them as one. Therefore, they remain constant.

Having the numbers for different time steps and domain sizes for LDJT, LJT, and JT, let us now identify the calculations LDJT performs for each time step to

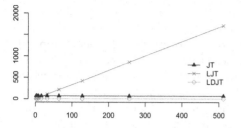

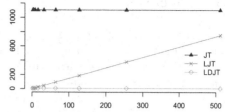

Fig. 10. (Par)clusters (y-axis) for different time steps (x-axis)

Fig. 11. (P)RVs in largest (par)cluster (y-axis) for different time steps (x-axis)

compare the calculations against LJT and JT. During message passing, LDJT computes $2 \cdot (3 - 1) = 4$ messages. For each message, LDJT sums out at most $6 - 1 = 5$ PRVs. Actually, LDJT sums out 1 PRV for each *inbound* message and 3 and 4 PRVs for the *outbound* messages. Additionally, for each time step LDJT needs to calculate an α_t message, for which LDJT needs to sum out all non-interface PRVs from the *out-cluster* parcluster. The additional efforts are similar to answering one additional query. In our case, the *out-cluster* parcluster contains 6 and the interface 3 PRVs. Therefore, LDJT sums out 3 PRVs to calculate an α_t message and overall computes 5 messages for each time step.

For a domain size of 32, JT has 70 clusters with maximal 1106 RVs in each of them. Hence, for every time step JT computes 138 messages and for every message needs to sum out 1105 RVs in the worst case. For the PDM unrolled for 64 time steps, LJT has 218 paclusters with at most 95 PRVs. Thus, LJT computes 434 messages and for every message needs to sum out 94 PRVs in the worst case. For 64 time steps, LDJT computes 320 messages, each with only a small friction of summing out operations. Further, only in case all evidence is known before for all 64 time steps, LJT computes 434 messages to be able to answer queries for all time steps. In case evidence is provided incrementally, LJT needs to perform message passing for each new evidence, which can result in 27776 messages. Therefore, accounting for the temporal and relational aspects of the model in LDJT significantly reduces the number of computations.

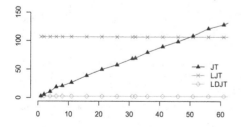

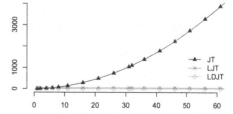

Fig. 12. (Par)clusters (y-axis) for different domain sizes (x-axis)

Fig. 13. (P)RVs in largest (par)cluster (y-axis) for different domain size (x-axis)

5 Conclusion

We present LDJT, a filtering and prediction algorithm for relational temporal models. LDJT answers multiple queries efficiently by reusing a compact FO jtree structure for multiple queries. Further, due to temporal m-separation, which is ensured by the *in-* and *out-clusters*, LDJT uses the same compact structure for all time steps $t > 0$. Furthermore, LDJT does not need to know the maximum number of time steps and allows for efficiently adding time steps dynamically. First results show that the number of computations LDJT saves compared to LJT and JT is significant.

We currently work on extending LDJT to also perform smoothing. Smoothing could also be helpful to deal with incrementally changing models. Other interesting future work includes a tailored automatic learning for PDMs, parallelisation as well as using local symmetries. Additionally, it would be interesting to include our work in PDBs, e.g., to handle correlated PDBs [8].

References

1. Ahmadi, B., Kersting, K., Mladenov, M., Natarajan, S.: Exploiting symmetries for scaling loopy belief propagation and relational training. Mach. Learn. **92**(1), 91–132 (2013)
2. Braun, T., Möller, R.: Lifted junction tree algorithm. In: Friedrich, G., Helmert, M., Wotawa, F. (eds.) KI 2016. LNCS (LNAI), vol. 9904, pp. 30–42. Springer, Cham (2016). https://doi.org/10.1007/978-3-319-46073-4_3
3. Braun, T., Möller, R.: Counting and conjunctive queries in the lifted junction tree algorithm. In: Croitoru, M., Marquis, P., Rudolph, S., Stapleton, G. (eds.) 5th International Workshop on Graph Structures for Knowledge Representation and Reasoning. LNCS, vol. 10775, pp. 54–72. Springer, Cham (2017). https://doi.org/10.1007/978-3-319-78102-0_3
4. Darwiche, A.: Modeling and Reasoning with Bayesian Networks. Cambridge University Press, Cambridge (2009)
5. Dignös, A., Böhlen, M.H., Gamper, J.: Temporal alignment. In: Proceedings of the 2012 ACM SIGMOD International Conference on Management of Data, pp. 433–444. ACM (2012)
6. Dylla, M., Miliaraki, I., Theobald, M.: A temporal-probabilistic database model for information extraction. Proc. VLDB Endow. **6**(14), 1810–1821 (2013)
7. Geier, T., Biundo, S.: Approximate online inference for dynamic Markov logic networks. In: Proceedings of the 23rd IEEE International Conference on Tools with Artificial Intelligence (ICTAI), pp. 764–768. IEEE (2011)
8. Kanagal, B., Deshpande, A.: Lineage processing over correlated probabilistic databases. In: Proceedings of the 2010 ACM SIGMOD International Conference on Management of data, pp. 675–686. ACM (2010)
9. Lauritzen, S.L., Spiegelhalter, D.J.: Local computations with probabilities on graphical structures and their application to expert systems. J. R. Stat. Soc. Ser. B (Methodol.) **50**, 157–224 (1988)
10. Manfredotti, C.E.: Modeling and Inference with Relational Dynamic Bayesian Networks. Ph.D. thesis, Ph.D. Dissertation, University of Milano-Bicocca (2009)

11. Milch, B., Zettlemoyer, L.S., Kersting, K., Haimes, M., Kaelbling, L.P.: Lifted probabilistic inference with counting formulas. In: Proceedings of AAAI, vol. 8, pp. 1062–1068 (2008)
12. Murphy, K., Weiss, Y.: The factored frontier algorithm for approximate inference in DBNs. In: Proceedings of the Seventeenth conference on Uncertainty in artificial intelligence, pp. 378–385. Morgan Kaufmann Publishers Inc. (2001)
13. Murphy, K.P.: Dynamic Bayesian Networks: Representation, Inference and Learning. Ph.D. thesis, University of California, Berkeley (2002)
14. Nitti, D., De Laet, T., De Raedt, L.: A particle filter for hybrid relational domains. In: Proceedings of the IEEE/RSJ International Conference on Intelligent Robots and Systems (IROS), pp. 2764–2771. IEEE (2013)
15. Papai, T., Kautz, H., Stefankovic, D.: Slice normalized dynamic Markov logic networks. In: Proceedings of the Advances in Neural Information Processing Systems, pp. 1907–1915 (2012)
16. Poole, D.: First-order probabilistic inference. In: Proceedings of IJCAI, vol. 3, pp. 985–991 (2003)
17. Richardson, M., Domingos, P.: Markov logic networks. Mach. Learn. **62**(1), 107–136 (2006)
18. de Salvo Braz, R.: Lifted First-Order Probabilistic Inference. Ph.D. thesis, Ph.D. Dissertation, University of Illinois at Urbana Champaign (2007)
19. Taghipour, N., Davis, J., Blockeel, H.: First-order decomposition trees. In: Proceedings of the Advances in Neural Information Processing Systems, pp. 1052–1060 (2013)
20. Taghipour, N., Fierens, D., Davis, J., Blockeel, H.: Lifted variable elimination: decoupling the operators from the constraint language. J. Artif. Intell. Res. **47**(1), 393–439 (2013)
21. Thon, I., Landwehr, N., De Raedt, L.: Stochastic relational processes: efficient inference and applications. Mach. Learn. **82**(2), 239–272 (2011)
22. Vlasselaer, J., Van den Broeck, G., Kimmig, A., Meert, W., De Raedt, L.: TP-Compilation for inference in probabilistic logic programs. Int. J. Approx. Reason. **78**, 15–32 (2016)
23. Vlasselaer, J., Meert, W., Van den Broeck, G., De Raedt, L.: Efficient probabilistic inference for dynamic relational models. In: AAAIWS'14-13 Proceedings of the 13th AAAI Conference on Statistical Relational AI, pp. 131–132. AAAI Press (2014)

Computer Human Interaction and Human Cognition

Defining Argumentation Attacks in Practice: An Experiment in Food Packaging Consumer Expectations

Bruno Yun[1], Rallou Thomopoulos[1,2(✉)], Pierre Bisquert[1,2],
and Madalina Croitoru[1]

[1] INRIA GraphIK, LIRMM (CNRS and Univ. Montpellier),
34392 Montpellier cedex 5, France
{bruno.yun,madalina.croitoru}@lirmm.fr
[2] INRA, IATE Joint Research Unit, 34060 Montpellier cedex 1, France
{rallou.thomopoulos,pierre.bisquert}@inra.fr

Abstract. In socio-economic systems, where actors are motivated by different objectives, interests and priorities, it is very difficult to meet all involved party expectations when proposing new solutions. Argumentative approaches have been proposed and demonstrated to be of added value when addressing such decision making problems. In this paper we focus on the following research question: "How to define an attack relation for argumentative decision making in socio-economic systems?" To address this question we propose three kinds of attacks that could be defined in the context of a precise application (packaging selection) and see how the non computer science experts evaluate, against a given set of decision tasks, each of these attacks.

1 Introduction

Socio-economic systems involve various actors who interact, while motivated by different objectives, interests and priorities. Food supply chains are examples of such complex systems, involving actors from producers to process and packaging industrials, distributors, recycling industry, etc. Conceiving sustainable food systems implies taking into account various kinds of concerns, including environmental issues (limited energy consumption, reduced GHG emissions, etc.), economic issues (limited costs for consumers, viable firms, source of employment, etc.), social issues (creating social link, ensuring good living conditions, etc.), ethical considerations (pursuing fairness and equity principles), sensorial preferences (appreciated taste and flavor), nutritional issues (contributing in healthy diets), sanitary issues (chemical and microbiological norms), and practical issues (shelf life, ease of use, etc.). As a consequence, these different concerns lead to inconsistent expectations. This raises the difficult question of how to best meet such expectations when proposing new solutions for the future.

Various methods of reasoning under such kind of inconsistency have already been proposed to tackle decision making in food supply chains. The main

© Springer International Publishing AG, part of Springer Nature 2018
P. Chapman et al. (Eds.): ICCS 2018, LNAI 10872, pp. 73–87, 2018.
https://doi.org/10.1007/978-3-319-91379-7_6

approaches employed by the state of the art are multi-criteria analysis or argumentation. In the rest of the paper we will focus on argumentation based approaches. This is justified by the fact that multi criteria design encounters a limitation which is particularly hard to address within mathematical' frameworks: striking the right balance between problem simplification and real-world complexity [21]. Furthermore, in the context of socio-economic systems decision making [9] the strength of argumentation lies in reasoning traceability, discussion fostering and decision explanation, all leading to a better acceptance of the final decision by all parties [6,20].

An argumentation system consists of a set of arguments and a binary relation on that set, expressing conflicts among arguments. Argumentation has been formalized both in philosophy and in computer science [14,18] and applied to decision making (e.g. [7,11]), deductive argumentation and defeasible logic programming [1,17] or for modelling different types of dialogues including negotiation or explanation (e.g. [3,4,13,19]). In the reminder of this paper we will focus on using argumentation for decision making. Originally, in order to capture a large class of problems, an argument [10] was seen as abstract entity. This abstraction poses problems when aiming to apply argumentation in practice for decision making. While an argument is intuitively understood as a statement for or against an action, the attack relation is much harder to discern. How is the attack relation obtained? Is it declared, deduced or computed? On which rationale is it defined? How should it be interpreted? Does it have a human-intuitive interpretation or a machine interpretation?

Against this background, in this paper we focus on the following research question: *"How to define an attack relation for argumentative decision making in socio-economic systems?"* Since the added value of argumentation lies in its principled interaction with the user we evaluate the potential answers to the above research question with the users. More precisely we propose three kinds of attacks that could be defined in the context of a precise application (packaging selection) and see how the non computer science experts evaluate, against a given set of decision tasks, each of these attacks.

The salient point of the paper lies in the fact that it represents, to the best knowledge of the authors, the first study in argumentation research that addresses the problem of attack modelling in practice. This is a significant problem since the structure of the argumentation graph is directly dictated by the attacks. Moreover, such structure plays a fundamental role in how difficult reasoning over the argumentation graph is [12,24].

The paper is structured as follows. After a recall on modelling choices in Sect. 2, the packaging case study is presented in Sect. 3. The experimental protocol of the paper is proposed in Sect. 4 and its results discussed in Sect. 5.

2   Argumentation Primer

Argumentation is a reasoning model based on the construction and evaluation of conflicting arguments [10]. An argumentation process follows three main steps:

(1) constructing arguments and counter-arguments via attacks, (2) evaluating the acceptability of the arguments using semantics defined on the resulting argumentation framework of step (1), and (3) obtaining the justified conclusions based on the set of acceptable arguments of step (2). These notions are formalised below (argumentation framework, acceptability semantics). Please note that for introducing acceptability semantics (in this paper we only considered the preferred semantics) we need to introduce three additional notions i.e. conflict-freeness, defence and admissibility.

Definition 1. *(Dung's argumentation framework). An argumentation framework is a pair $AF = (\mathcal{A}, \mathcal{R})$ where $\mathcal{A}$ is a set of arguments and $\mathcal{R} \subseteq \mathcal{A} \times \mathcal{A}$ is an attack relation. An argument a attacks an argument a' if and only if $(a, a') \in \mathcal{R}$.*

Definition 2. *(Conflict-free, Defense, Admissibility). Let an argumentation framework $AF = (\mathcal{A}, \mathcal{R})$ and $B \subseteq \mathcal{A}$. Then:*

- *B is conflict-free if and only if $\nexists a_i, a_j \in B$ such that $(a_i, a_j) \in \mathcal{R}$;*
- *B defends an argument $a_i \in B$ if and only if for each argument $a_j \in \mathcal{A}$, if $(a_j, a_i) \in \mathcal{R}$, then $\exists a_k \in B$ such that $(a_k, a_j) \in \mathcal{R}$;*
- *a conflict-free set B of arguments is admissible if and only if B defends all its elements.*

A preferred extension is a maximal (with respect to set inclusion) admissible set of arguments.

In the above framework, an argument is abstract and can be differently instantiated according to various contexts [22]. In general, an argument gives a reason for believing a claim, or for doing an action and it is commonly seen as a set of statements composed of one (or more) premise(s) and a conclusion linked by a logical relation. In decision making, arguments can be intuitively understood as being statements to support, contradict, or explain opinions or decisions [2]. More precisely, in decisional argumentation frameworks [16], the argument definition is provided with additional features: the decision (also referred to as action, option or alternative) and the goal (also referred to as target). In some studies arguments are also associated with specific actors. An application of a decision-oriented argumentation framework to a real-life problem concerning food policy can be found in [8], where a recommendation regarding the provision of whole-grain bread was analyzed a posteriori. In that case, each argument is associated with the action it supports. Based on the above rationale, in this paper, the structure for argument modelling is defined as a tuple composed of a target (a goal), a conclusion inferred from the target choice (i.e. considering/or not this target will imply that the following condition will hold), an aligning of the argument with respect to the target (i.e. if the condition holds in the presence or absence of the target) and a priority rank. Formally:

Definition 3. *An argument is a tuple $a = (T, C, A, R)$ where:*

- T *is the target of the argument;*
- C *is the conclusion of the argument;*
- $A \in \{for, against\}$ *is the alignment of the argument (i.e. for or against);*
- $R \in \mathbf{N}$ *is the rank of the argument.*

For any argument a, we denote by *Target(a)* the target of the argument, *Conclusion(a)* the conclusion of the argument a, *Alignment(a)* the alignment of a and *Rank(a)* the rank of the argument. In Table 1 the set of arguments considered in our experimentation is given for illustrative purposes. More details about the use case scenario and the experimentation protocol will be provided in the next section.

Table 1. List of arguments of the experimentation use case.

Name	Target	Conclusion	Alignment	Rank
$a1$	WC	$shock\_resistant$	for	14
$a3$	WC	$taste\_preservation$	for	8
$a5$	WC	$can\_smell\_product$	for	1
$a7$	WC	$can\_see\_product$	for	9
$a11$	WC	$incites\_to\_eat$	for	4
$a14$	WC	$ambiant\_preservation$	for	18
$a15$	WC	$refrigerator\_preservation$	for	19
$a16$	WC	$have\_aerations$	for	3
$a19$	WC	$protect\_environment$	for	6
$a20$	WC	$reusable$	for	13
$a21$	WC	$harmful\_effect$	$against$	22
$a2$	PRL	$shock\_resistant$	for	15
$a4$	PRL	$taste\_preservation$	for	20
$a8$	PRL	$can\_see\_product$	for	5
$a12$	PRL	$incites\_to\_eat$	for	16
$a17$	PRL	$have\_aerations$	for	17
$a9$	PPF	$can\_see\_product$	for	10
$a22$	PPF	$NOT\_reusable$	$against$	21
$a6$	OPC	$can\_smell\_product$	for	7
$a10$	OPC	$can\_see\_product$	for	2
$a13$	OPC	$incites\_to\_eat$	for	12
$a18$	OPC	$have\_aerations$	for	11

In structured argumentation (e.g. logic based argumentation frameworks where arguments are obtained as instantiations over an inconsistent knowledge base) three kinds of attacks have been defined: undercut, rebut and undermine.

[5]. The intuition of these attacks is either to counter the premise of the opposing argument (the undercut), the conclusion (the rebut) or the logical step that allowed the inference between premise and conclusion (undermine). In abstract argumentation the set of attacks is simply considered as given. A particular aspect to be considered is when the argumentation framework is enhanced with a set of preferences (in this paper the preferences over arguments are cardinal and expressed as ranks). Classically, when preferences are present the attack relation can be modified in order to take into account the induced precedence.

In this paper, since the considered definition of the attack places our work between structured and abstract argumentation (i.e. we impose some structure on the argument given the decision task at hand but do not logically formalise the target and the conclusion), we need to decide how the attack is defined. Following the above intuition, attacking an argument could be achieved by raising doubts about its acceptability and (i) questioning its target, (ii) questioning the alignment with respect to the target, or (iii) using the argument rank. Formally, we consider the following three attack relations $\mathcal{R}_1$, $\mathcal{R}_2$ and $\mathcal{R}_3$:

Definition 4. *Let $a, b \in \mathcal{A}$ be two arguments. We say that:*

- *$(a, b) \in \mathcal{R}_1$ iff $Target(a) \neq Target(b)$;*
- *$(a, b) \in \mathcal{R}_2$ iff $Target(a) = Target(b)$ and $Alignment(a) \neq Alignment(b)$;*
- *Let $\succeq$ be a partial order on arguments, $(a, b) \in \mathcal{R}_3$ iff $Target(a) \neq Target(b)$, $Alignment(a) = Alignment(b) = for$ and $Rank(a) \leq Rank(b)$.*

Let us consider again the arguments of Table 1. We have that $(a1, a2) \in \mathcal{R}_1$ since $Target(a1) = WC$ and $Target(a2) = PRL$. Furthermore, we have that $(a1, a21) \in \mathcal{R}_2$ since $Target(a1) = Target(a21) = WC$ and $Alignment(a1) = for$ and $Alignment(a21) = against$. Finally, we have that $(a1, a2) \in \mathcal{R}_3$ since $Rank(a1) \leq Rank(a2)$.

All of the 3 attacks can be a priori justified from examples as being intuitive, and have been in the literature (see e.g. [8,23]). Our objective here is to propose an a posteriori and experimental evaluation, which has not been proposed. In this evaluation, we will consider several criteria: (i) adequation to human spontaneous way of reasoning; (ii) human interpretability of the results; (iii) added-value of computarization for the human. Adequation to human spontaneous way of reasoning will be used as the prime criterion to assert intuitiveness.

3   Case Study

In the framework of the Pack4Fresh agri-food research project supported by the INRA and CIRAD research institutes, a study was launched to design optimized innovative food packagings.

Food packagings play an crucial part in the food market, since they perform multiple functions: marketing appeal, information about the provenance of the products, nutrition facts, food preservation as long as possible associated with food waste reduction, logistic practicality such as the possibility to stack the products for instance. However, food packagings also have harmful effects, in particular on the environment, since it generates waste material and requires energy-intensive manufacturing, etc. At the present time, active research is ongoing to design and develop new-generation, biobased, biodegradable, "intelligent" food packagings. It is mainly focused on technical aspects of packaging, such as properties of the materials used, matter flows through the packaging material, etc. Yet, to be acceptable and usable, these new-generation packagings have to take into account the characteristics expected or appreciated by the users at all levels of the food supply chain (production, transportation, distribution, storage, consumption, etc.). Hence the importance of explicitly identifying these expectations, as exhaustively as possible, starting from cases of (i) existing packagings and (ii) simple systems, excluding in particular labelling information and advanced preservation properties such as modified atmosphere.

Pursuing this objective, in this study it was decided to focus on the case of strawberry packagings, because of the local accessibility of this product and its properties as a perishable product, subject to much wastage by consumers due to bad habits. The following four existing alternatives were considered and the aim of the study was to best identify and analyze the pros and cons of each of them:

– A opened plastic container without lid or plastic film (OPC).
– A opened wooden container without lid or plastic film (WC).
– A plastic container with a rigid lid (PRL).
– A plastic container with a plastic film (PPF).

In order to determine the viewpoints and needs of the wider class of consumers w.r.t. the four previous packagings, a web survey was developed. The survey was given to a representative (with regard to age and socio-professional category) sample of 840 persons. As an example, the questions asked by the survey were of the form: "*Do you think that this packaging has a harmful effect on the strawberries?*". The possible answers were "Yes", "No" or "Neutral".

In a first step, a pretreatment of neutral answers was performed. Questions with an important part of neutral answers were eliminated and not exploited at all in the rest of the study. Therefore, we considered the ratio of neutral answers should be less than in the case where answers would have been obtained randomly, i.e. less than $1/3$.

Then in a second step we extracted statements from the survey representing the general viewpoint of the consumers w.r.t. the packaging as follows. For a given question, let n_1 be the number of "Yes" answers and n_2 the number of "No" answers. If the ratio $n_1/(n_1 + n_2)$ is superior to a given threshold α, a positive statement is extracted. Conversely, if the ratio $n_2/(n_1 + n_2)$ is superior to α, a negative statement is extracted. Otherwise no statement was extracted. The higher the ratio, the prior the rank of the statement is. Again we set α at $2/3$ to ensure the results are statistically different from answers that would have been obtained randomly (i.e. $\alpha \approx 0.5$). For instance, for the given assertion of whether "the open wooden containers are recyclable", the number of "Yes" answers of the survey participants is 658, whereas the number of "No" answers is 45. We have $658/(658 + 45) > 2/3$, thus a positive statement for the open wooden container in the form: "The consumers think that the opened wooden container is recyclable" is extracted. The same question concerning the plastic container with the plastic film provided 165 "Yes" answers and 370 "No" answers. Since we have $370/(165 + 370) > 2/3$, a negative statement against the plastic container with the plastic film, of the form: "The consumers think that the plastic container with the plastic film is not recyclable", is extracted.

These statements were transformed into arguments as shown in Table 1. The different kinds of attacks as defined in the previous section were computed. The resulting three argumentation frameworks (sharing the set of arguments) were the inputs for the experimentation protocol described in the next section.

4 Experimentation Protocol

The experimentation was composed of 3 groups (A, B and C) of 7, 8 and 10 persons respectively, all experts in agri-food research but not necessarily experts in packaging conception. Each group rigorously followed the experimentation's framework described in Fig. 1. The difference in the group sizes is an experimental constraint due to logistic reasons. Although reduced, the experiment size is conform to ratio requirements [15] which recommend a minimum number of observations (here participants) equivalent to the number of entries (here arguments).

These three groups were first shown a textual description of the use-case. Then, they were asked to grade the 4 packagings by giving them a score between 1 and 4 (see Fig. 2). In Fig. 2, the blue bar above WC_A means that in group A, there were 4 persons that gave the score 1 to the wooden container whereas the grey bar above PPF_C means that in group C, 8 persons gave the score 4 to the plastic container with plastic film. Note that 1 is considered the best score and 4 the worst. The sum may be superior to the number of participants because

The aggregated score of a packaging $p \in \{OPC, WC, PRL, PPF\}$ was computed as $Score(p) = \sum_{i\in\{1,2,3,4\}} NumberPersons(p,i) * i$.

In the formula above, $NumberPersons(p,i)$ represents the number of persons that gave the score i to the packaging p. Please note the smaller the score of a

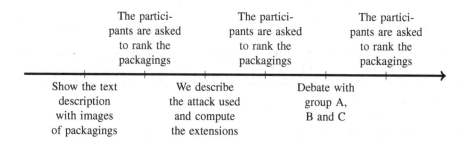

Fig. 1. Timeline of the experiment conducted

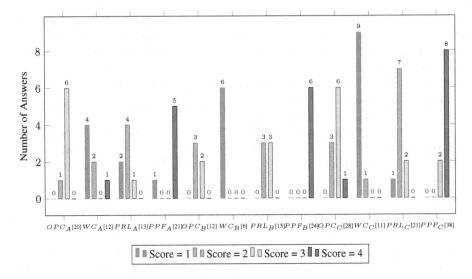

Fig. 2. Scoring for each group before the argumentation methods

packaging, the better that packaging is. In Fig. 2, the aggregated score of each packaging in each group is enclosed in square brackets. The detailed analysis of the results is as follows:

- In group A, the WC is considered the "best" packaging (score 12), the PRL is second (score 13). It is followed by the OPC (score 20) and the PPF at the last place (score 21).
- In group B, the WC is obviously the "best" packaging (score 6), it is followed by the OPC (score 12) and the PRL (score 15). The all agree to give the PPF the last place (score 24).
- In group C, the WC is the "best" packaging (score 11), the PRL is second (score 21). It is followed by the OPC (score 28) and the PPF is at the last place (score 38).

From the above we can conclude that, before the experiment, the three groups relatively agree on the ranking of packagings: $WC > PRL > OPC > PPF$ (for groups A and C) and $WC > OPC > PRL > PPF$ (for group B).

The experiment we held out for each group was as follows:

- For group A, we considered the argumentation graph $\mathcal{AS} = (\mathcal{A}, \mathcal{R})$ where $\mathcal{A}$ is the set of arguments of Table 1 with $Alignment(a) = for$ for each $a \in \mathcal{A}$ and $\mathcal{R} = \mathcal{R}_1$. They were shown the following list of preferred extensions:
  - $\varepsilon_{PPF} = \{a9\}$
  - $\varepsilon_{OPC} = \{a6, a10, a13, a18\}$
  - $\varepsilon_{PRL} = \{a2, a4, a8, a12, a17\}$
  - $\varepsilon_{WC} = \{a1, a3, a5, a7, a11, a14, a15, a16, a19, a20\}$

In this case, one preferred extension of the argumentation graph corresponds to one packaging and contains all the arguments that are in favor of this option.

- For group B, we considered the argumentation graph $\mathcal{AS} = (\mathcal{A}, \mathcal{R})$ where $\mathcal{A}$ is the set of arguments of Table 1 and $\mathcal{R} = \mathcal{R}_2$. They were shown the following list of preferred extensions:

  - $\varepsilon_1 = \{a2, a4, a6, a8, a10, a12, a13, a17, a18, a21, a22\}$
  - $\varepsilon_2 = \{a2, a4, a6, a8, a9, a10, a12, a13, a17, a18, a21\}$
  - $\varepsilon_3 = \{a1, a2, a3, a4, a5, a6, a7, a8, a9, a10, a11, a12, a13, a14, a15, a16, a17, a18, a19, a20\}$
  - $\varepsilon_4 = \{a1, a2, a3, a4, a5, a6, a7, a8, a10, a11, a12, a13, a14, a15, a16, a17, a18, a19, a20, a22\}$

  In this case, one extension cannot contain the arguments for and against one packaging at the same time. Here, ε_1 contains the arguments for PRL, the arguments for OPC, the arguments against WC and the arguments against PPF, ε_2 contains the arguments for PRL, the arguments for OPC, the arguments for PPF and the arguments against WC, ε_3 contains all the arguments for a packaging and ε_4 contains the arguments for PRL, the arguments for OPC, the arguments for WC and the arguments against PPF.

- For group C, we considered the argumentation graph $\mathcal{AS} = (\mathcal{A}, \mathcal{R})$ where $\mathcal{A}$ is the set of arguments of Table 1 with $Alignment(a) = for$ for each $a \in \mathcal{A}$ and $\mathcal{R} = \mathcal{R}_3$. They were shown the following list of preferred extensions $\varepsilon_{WC} = \{a1, a3, a5, a7, a11, a14, a15, a16, a19, a20\}$, that contains all the arguments for the wooden container.

After conducting the argumentation experiment and showing them the resulting extensions, the participants were asked again to grade the four packagings by giving them a score between 1 and 4 (see Fig. 3). Aggregated scores are given in square brackets and the detailed analysis of the results is as followed:

- In group A, the WC is considered the "best" packaging (score 7), the PRL is second (score 14), third is the OPC (score 21) and the PPF (score 28) is at the last place. Interesting enough, every participant agreed to this ranking.
- In group B, the OPC is the "best" packaging (score 8), it is followed by the PRL (score 10) and WC get the third place (score 16). The PPF is ranked last (score 22).
- In group C, the WC is the "best" packaging (score 8), the OPC is second (score 18). It is followed by the PRL (score 20) and the PPF is at the last place (score 27).

Once the scoring of packagings was completed, we reunited the three groups (A, B and C) and proceeded with a group debate. Each group explained its approach and revealed their rankings on packagings. We then asked the three groups to grade again the four packagings by giving them a score between 1 and 4 (see Fig. 4). Aggregated scores are given in square brackets and the detailed analysis of the results is as followed:

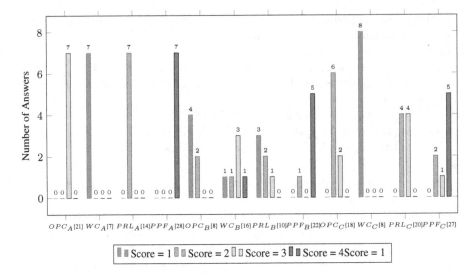

Fig. 3. Scoring for each group after the argumentation methods

- In group A, the WC is considered the "best" packaging (score 6), the PRL is second (score 12), third is the OPC (score 18) and the PPF is at the last place (score 24).
- In group B, the WC is the "best" packaging (score 9), it is followed by the OPC (score 19) and PRL at the second/third place (score 19). The PPF is ranked last (score 24).
- In group C, the WC is the "best" packaging (score 10), the PRL is second (score 23). It is followed by the OPC (score 25) and the PPF is at the last place (score 35).

5   Discussion

Let us summarise the different rankings expressed in this paper in Table 2.

Table 2. Overview of the results

Group	Before the experiment	After experiment	After debate
A	$WC > PRL > OPC > PPF$	$WC > PRL > OPC > PPF$	$WC > PRL > OPC > PPF$
B	$WC > OPC > PRL > PPF$	$OPC > PRL > WC > PPF$	$WC > OPC \sim PRL > PPF$
C	$WC > PRL > OPC > PPF$	$WC > OPC > PRL > PPF$	$WC > PRL > OPC > PPF$

From this table we can conclude that the argumentation method of group A does not change the ranking of packagings. The argumentation method of

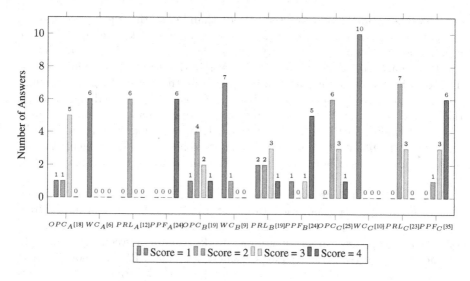

Fig. 4. Scoring for each group after the debate

group B seems to be misleading as we can see great changes in the ranking after the experiment which are then revised after the debate. It seems like the argumentation method of group C has some minor effect on the ranking since $PRL > OPC$ before the experiment and $OPC > PRL$ after. One interesting result is that the ranking on packagings is similar before the experiment and after the debate. To go further, let us consider in more detail the ranking statistics as well as the individual comments obtained in each group.

If we examine the statistics of Figs. 2, 3 and 4, we can make the following observations. In Fig. 2, at the beginning of the experiment before applying the argumentation methods, the global ranking (mixing the answers of the three groups) is clearly $WC > PRL > OPC > PPF$, however the three groups show some nuances in their profiles. Group A shows a clear view about which 2 packagings should be ranked third and last, but seems to be hesitating about how to discriminate between WC and PPF for ranks first and second. Group B shows a clear view of which packaging should be ranked first and last (respectively WC and PPF) but seems more doubtful on the discrimination between the intermediate rankings (PRL or OPC), since the ranking between PRL and OPC is close to ex aequo in this group. Group C offers a quite categoric ranking with little exceptions with $WC > PRL > OPC > PPF$. Since all participants have the same information, the variations observed between the three groups are due to individual differences in interpretation.

In Fig. 3, after the experiment applying the different argumentation methods, the differences in the profiles of the three groups become much deeper. Group A provides an answer with no ambiguity with the unanimous ranking $WC > PRL > OPC > PPF$. Thus, this group seems to have gained confidence in their initial answer by applying the argumentation method. Group B, on the contrary, has a much more confused opinion than before the experiment. Indeed, its ranking number one has changed in favour of the packagings that have no negative arguments (OPC and PRL). Its ranking number 2 is now distributed on all the packaging options. WC, which was initially ranked first, has become number 3 because of its negative argument. Only the last ranking, PPF, which also has a negative argument, is unchanged. Group C has gained confidence in their initial first rank WC, since 100% of the participants now rank WC first. Their last ranking, PPF, is also confirmed. However ranks 2 and 3 tend to be inversed, with OPC coming before PRL. This can be explained by the prior ranks of the arguments associated with OPC, these ranks being further examined by this group.

In Fig. 4, after the debate common to the three groups, the final opinions tend to be more consensual, each group retrieving additional information from the other two experimentations. Hence, the first rank for WC and the last rank for PPF are majoritary in all the groups. Ranks 2 and 3 between OPC and PRL are more controversial, taking into consideration that PRL has more positive arguments, that none of them has negative arguments, but that OPC has prior-ranked arguments.

Moreover, comments given by the participants allow to provide complementary analysis. At the beginning of the experiment before applying the argumentation methods, the comments show that the participants intuitively perform the ranking by seeking a balance between a great number of positive arguments and restricted occurrences of negative arguments, e.g.: *"The wooden packaging has lots of positive arguments but also a strong negative argument since it is harmful to strawberries. Thus, the plastic packaging with a rigid lid comes first since it does not have any negative arguments."*. *"There is a lot of positive arguments and no negative arguments for the plastic container with a rigid lid; The wooden container seems to be ranked second despite of having more positive arguments because of the presence of negative arguments (possibility to smell, harmful effects, etc.)"*. *"I compare the options w.r.t. the number of positive and negative arguments."*.

After the experiment, each groups provides interpretations according to the method used. In group A the ranking is reduced to a counting method: *"We sort according to the number of arguments. There is a bigger extension for the wooden container."*. Group B expresses the difficulty in interpreting the extensions which mix arguments from all packaging options: *"There is no possible choice because of the large number of extensions. However, the plastic open container and the plastic container with a rigid lid appear in all the 4 extensions which means that we can prioritise them."*. Group C identifies the method is more suitable to compute the best choice than to rank the options, since the output contains only one extension here: *"The only extension available. The answers for the wooden container are stronger, there is no knowledge on the other packagings"*.

Based on the observations above, we can conclude on the strong and weak points of each attack definition:

- by separating the positive arguments associated with a given option into a distinguished extension, $\mathcal{R}_1$ allows to rank the options by simply counting the number of arguments in each extension, which is what the participants intuitively perform before the experiment. However, $\mathcal{R}_1$ does not take into account negative arguments, whereas the participants intuitively do;
- $\mathcal{R}_2$ takes into account the bipolar aspect of decision making by considering both positive and negative arguments, as participants intuitively do. But having positive and negative arguments for a given option in separate extensions, and even more mixing arguments from different options in the same extension, is misleading and practically impossible to interprete by the participants;
- the added-value of $\mathcal{R}_3$ is to take into account and highlight the prioritization of arguments, which at first glance is not natural to the participants and thus useful to computerize. However the non-symmetric attack relation it provides does not guarantee to keep one extension per option, which is relevant for the selection of best alternatives but leads to a lack of knowledge for the participants about the ranking of the other alternatives.

6   Conclusion

In this paper we have introduced an experimental study of a packaging conception use case for strawberries. The aim of the experiment was to investigate how the different attack relations behave on a decision making argumentation framework used by non computer science experts. As seen from our results we can infer that the $\mathcal{R}_1$ attack showed to be the most intuitive and easy to use one, whereas $\mathcal{R}_2$ was the closest to the bipolar way of reasoning naturally adopted by the participants by considering both positive and negative arguments, and $\mathcal{R}_3$ seemed to provide the most added-value in computerizing the approach by performing computations with priority ranks. [23] may be cited as an attempt to combine several of these advantages through a preference-based ranking approach. From the experiment analysed in the present paper, the decision mechanism could be done by an algorithm in order to simulate human experts taking into account attack relations. Obviously from our results, such an algorithm should be primarily based on $\mathcal{R}_1$ and possibly refined using $\mathcal{R}_3$ then $\mathcal{R}_2$.

Let us conclude this paper with one last remark. Further data are available for analysis in the context of this use case as the consumer survey was complemented by interviews with scientists experts of packaging technologies and with actors of the distribution industry (wholesalers, quality managers, etc.). While in this paper we focus on consumers' answers, in order to study the intuitiveness impact of the attack definition in a delimitated case involving only one category of stakeholders, integrating the various stakeholder opinions is ongoing current work.

Acknowledgements. The authors acknowledge the participants to the ECIDCM 2016 training school. We also acknowledge the support of the Pack4Fresh project. Many thanks to Patrice Buche for his insights and help with the experimental set up.

References

1. Amgoud, L., Bodenstaff, L., Caminada, M., McBurney, P., Parsons, S., Prakken, H., Veenen, J., Vreeswijk, G.: Final review and report on formal argumentation system. deliverable d2.6 aspic. Technical report (2006)
2. Amgoud, L., Prade, H.: Using arguments for making and explaining decisions. Artif. Intell. **173**(3–4), 413–436 (2009)
3. Arioua, A., Croitoru, M.: Formalizing explanatory dialogues. In: Beierle, C., Dekhtyar, A. (eds.) SUM 2015. LNCS (LNAI), vol. 9310, pp. 282–297. Springer, Cham (2015). https://doi.org/10.1007/978-3-319-23540-0_19
4. Arioua, A., Croitoru, M.: A dialectical proof theory for universal acceptance in coherent logic-based argumentation frameworks. In: ECAI, pp. 55–63 (2016)
5. Besnard, P., Hunter, A.: Elements of Argumentation. MIT Press, Cambridge (2008)
6. Bisquert, P., Croitoru, M., de Saint-Cyr, F.D., Hecham, A.: Formalizing cognitive acceptance of arguments: durum wheat selection interdisciplinary study. Mind. Mach. **27**(1), 233–252 (2017)
7. Bonet, B., Geffner, H.: Arguing for decisions: a qualitative model of decision making. In: Horvitz, E., Jensen, F. (eds.) 12th Conference on Uncertainty in Artificial Intelligence, Portland, pp. 98–105. Morgan Kaufmann (1996)
8. Bourguet, J.-R., Thomopoulos, R., Mugnier, M.-L., Abécassis, J.: An artificial intelligence-based approach to deal with argumentation applied to food quality in a public health policy. Expert Syst. Appl. **40**(11), 4539–4546 (2013)
9. Delhomme, B., Taillandier, F., Abi-Zeid, I., Thomopoulos, R., Baudrit, C., Mora, L.: Designing an argumentative decision-aiding tool for urban planning. In: OPDE 2017, Montpellier, France, October 2017
10. Dung, P.M.: On the acceptability of arguments and its fundamental role in non-monotonic reasoning, logic programming and n-person games. Artif. Intell. J. **77**, 321–357 (1995)
11. Fox, J., Das, S.K.: Safe and Sound - Artificial Intelligence in Hazardous Applications. MIT Press, Cambridge (2000)
12. Gaggl, S.A., Linsbichler, T., Maratea, M., Woltran, S.: Benchmark selection at ICCMA 2017 (2017)
13. Kraus, S., Sycara, K.P., Evenchik, A.: Reaching agreements through argumentation: a logical model and implementation. Artif. Intell. **104**(1–2), 1–69 (1998)
14. Mackenzie, J.: Begging the question in non-cumulative systems. J. Philos. Logic **8**, 117–133 (1979)
15. Marugán, A.P., Márquez, F.P.G.: Decision-Making Management. Academic Press, Cambridge (2017)
16. Ouerdane, W., Maudet, N., Tsoukiàs, A.: Argumentation theory and decision aiding. In: Ehrgott, M., Figueira, J., Greco, S. (eds.) Trends in Multiple Criteria Decision Analysis, vol. 142. Springer, Boston (2010). https://doi.org/10.1007/978-1-4419-5904-1_7
17. Prakken, H.: An abstract framework for argumentation with structured arguments. Argum. Comput. **1**(2), 93–124 (2011)
18. Rescher, N.: The role of rhetoric in rational argumentation. Argumentation **12**(2), 315–323 (1997)
19. Sycara, K.P.: Persuasive argumentation in negotiation. Theor. Decis. **28**(3), 203–242 (1990)
20. Thomopoulos, R., Croitoru, M., Tamani, N.: Decision support for agri-food chains: a reverse engineering argumentation-based approach. Ecol. Inform. **26**(2), 182–191 (2015)

21. Tremblay, J., Abi-Zeid, I.: Value-based argumentation for policy decision analysis: methodology and an exploratory case study of a hydroelectric project in Québec. Ann. Oper. Res. **236**(1), 233–253 (2016)

22. Walton, D., Macagno, F.: A classification system for argumentation schemes. Argum. Comput. **6**(3), 219–245 (2015)

23. Yun, B., Bisquert, P., Buche, P., Croitoru, M.: Arguing about end-of-life of packagings: preferences to the rescue. In: Garoufallou, E., Subirats Coll, I., Stellato, A., Greenberg, J. (eds.) MTSR 2016. CCIS, vol. 672, pp. 119–131. Springer, Cham (2016). https://doi.org/10.1007/978-3-319-49157-8_10

24. Yun, B., Vesic, S., Croitoru, M., Bisquert, P., Thomopoulos, R.: A structural benchmark for logical argumentation frameworks. In: Adams, N., Tucker, A., Weston, D. (eds.) IDA 2017. LNCS, vol. 10584, pp. 334–346. Springer, Cham (2017). https://doi.org/10.1007/978-3-319-68765-0_28

Empirically Evaluating the Similarity Model of Geist, Lengnink and Wille

Moritz Schubert and Dominik Endres[(✉)]

Philipps University Marburg, Marburg, Germany
moritz.schubert@students.uni-marburg.de, dominik.endres@uni-marburg.de

Abstract. In applications of formal concept analysis to real-world data, it is often necessary to model a reduced set of attributes to keep the resulting concept lattices from growing unmanageably big. If the results of the modeling are to be used by humans, e.g. in search engines, then it is important that the similarity assessment matches human expectations. We therefore investigated experimentally if the set-theoretic reformulation of Tversky's contrast model by Geist, Lengnink and Wille provides such a match. Predicted comparability and its direction was reflected in the human data. However, the model rated a much larger proportion of pairs as incomparable than human participants did, indicating a need for a refined similarity model.

1 Introduction

Similarity is highly relevant in the field of concept-based data mining. The lattices produced by formal concept analysis (FCA) [2] on raw real-world data are frequently too large or noisy to be useful. Approaches for the reduction of such lattices, such as iceberg lattices [9] or conceptual stability [4], are based on the data statistics. We argue here that such reduction techniques should also be based on *human expectations*, if the results are to be digested by human users. One key aim of FCA is to make the relational structure of the data explicit. It is therefore important that relevant relations, e.g. similarity as perceived by humans, are preserved as well as possible in the reduction process.

Similarity is also a crucial construct in the psychological study of concepts. Two of the most influential theories in the field, the prototype and the exemplar view, rely heavily on similarity when explaining the formation and structure of cognitive concepts [5].

Tversky [10] developed the contrast model to formalize key properties of human similarity perception. This model implies a total similarity ordering of continuously many objects. However, an infinite object space for a theory about human similarity judgment is highly implausible, because the capacity of the human brain is finite. Geist et al. [3] published a modification of Tversky's model, hereafter the 'GLW model', where they dropped the questionable requirements, yielding a partial order for similarities. In this paper, we present the first (to our knowledge) empirical test of this GLW model on human participants.

© Springer International Publishing AG, part of Springer Nature 2018
P. Chapman et al. (Eds.): ICCS 2018, LNAI 10872, pp. 88–95, 2018.
https://doi.org/10.1007/978-3-319-91379-7_7

In Sect. 2 we will introduce the theoretical models pertinent to our study. We present the experimental design and results in Sects. 3.1 and 3.2: the GLW model matches human expectations well whenever it predicts similarity comparability, but not otherwise. We discuss these results and possible implications for future models in Sect. 4.

2   Similarity Models

A large number of models for quantifying similarity have been proposed in the literature. Most popular is the distance model [7,8]. In the following, assume that we have a set G of objects $a, b \in G$, and let A, B be the representations of a, b. The distance model assumes that (dis)similarity ratings can be described as distances $d(A, B)$ between points A, B in an Euclidean space. Thus human similarity assessment needs to obey the usual distance axioms, in particular the triangle inequality $d(A, B) + d(B, C) \geq d(A, C)$ and symmetry $d(A, B) = d(B, A)$.

2.1   Contrast Model

Tversky [10] criticized the assumption that the triangle inequality and symmetry hold for perceived similarity. Experimental evidence indicates that humans rate non-salient stimuli more similar to salient ones than vice versa, violating symmetry.

In agreement with [10] we also object to the triangle inequality. Consider this counterexample: Let $c, t, b \in G$ be a yellow chick, yellow tennis ball and basketball, respectively. c is similar to t (because of their common colour) and t is similar to b (because of their common shape), however c is very dissimilar to b. If distances were suitable dissimilarity representations, we would therefore expect $d(c, b) > d(c, t) + d(t, b)$, which contradicts the triangle inequality.

Taken together, these arguments show that distance is not a suitable construct to describe human similarity assessment. Therefore Tversky [10] proposed a new model that is based on sets of attributes as descriptions of objects. He argued that sets were more natural representations of objects than points in a metric space. His *contrast model* is based on the following

Axioms. Let G be the set of all relevant objects, M be the set of all attributes of the objects $g \in G$ and $I \subseteq G \times M$ the incidence relation, i.e. $(g, m) \in I$ indicates that object g has attribute m. For brevity, we will denote with A, B, C the attribute sets of the objects a, b, c, e.g. $A = \{m | m \in M \wedge (a, m) \in I\}$. The similarity $s(a, b)$ between a and b is partially ordered. Let $\overline{A} = \{m | m \in M \wedge m \notin A\}$ be the complement of A. The contrast model is based on the following five axioms:

1. **Matching:**
$$s(a, b) = F(A \cap B, A \cap \overline{B}, B \cap \overline{A}) \tag{1}$$
i.e. similarity is a *matching function* $F()$ of shared and separating attributes.

2. **Monotonicity:** Similarity increases with common attributes, and decreases with separating ones:

$$s(a, b) \geq s(a, c) \Leftrightarrow A \cap B \supseteq A \cap C \wedge A \cap \overline{B} \subseteq A \cap \overline{C} \wedge B \cap \overline{A} \subseteq C \cap \overline{A} \tag{2}$$

3. **Solvability:** stipulates that for every real t in a similarity interval between $s(a, b) > t > s(c, d)$, there must be an object pair (e, f) with similarity t. This implies a continuous number of object pairs, and therefore continuously many objects.

The axioms of *independence* and *invariance*, which are concerned with the factorial structure of $F()$ and the equivalence of attribute intervals, are not important for our considerations in the following. Suffice to say that in conjunction, these axioms allow for a proof of the

Representation Theorem. If those five axioms hold then it can be shown [10], that there exists a similarity scale S and a non-negative scale f such that for all $a, b, c, d \in G$:

$$S(a, b) \geq S(c, d) \text{ iff } s(a, b) \geq s(c, d) \tag{3}$$

$$S(a, b) = \theta f(A \cap B) - \alpha f(A \cap \overline{B}) - \beta f(B \cap \overline{A}), \text{ for some } \theta, \alpha, \beta \geq 0 \tag{4}$$

f and S are interval scales, hence similarity is totally ordered.

Tversky calls the model "contrast model", because the commonalities of two objects are *contrasted* with their differences.

Critique of the Contrast Model. A consequence of the representation theorem is that similarities are totally ordered: one should always be able to determine whether object pair (a, b) is more, less or equally similar than/to object pair (c, d). The axiom of solvability is introduced specifically to establish a total order. However, this axiom is problematic: solvability requires continuously many objects. An infinite object space for a theory about human similarity judgment is highly implausible, because the capacity of the human brain is finite. Additionally, *solvability* posits the existence of new object pairs that lie on a proposed continuum between other pairs of objects (see [10]). For example, if a human knows about tomatoes and cars (both having round features), they should also know objects that are half car and half tomato. The existence of such "transitional" objects seems unlikely to us. Consequently, we disagree with Tversky's conclusion that similarity scales are total orders. For example, when asked whether a car is more similar to a tomato than a bicycle is to a potato, most people would probably say that this a silly comparison to make. A useful similarity model should capture this notion. One order-theoretic construct that might be suitable for this purpose is *incomparability*, i.e. similarity should be formalized as a partial order.

Partial ordering is at the core of the GLW model [3], which we will describe next.

2.2   GLW Model

The stated goal of Lengnink, Geist and Wille [3] was to generalize Tversky's [10] contrast model.

Axioms. The objects $a, b, c, d \in G$ are characterized by their attribute sets $A, B, C, D \in \mathcal{P}(M)$, where $\mathcal{P}(M)$ denotes the power set of M.

The elements $(A, B), (C, D) \in \mathcal{P}(M)^2$ are ordered in the following way:

$$(A, B) \geq (C, D) :\Leftrightarrow A \cap B \supseteq C \cap D \wedge \tag{5}$$

$$A \cap \overline{B} \subseteq C \cap \overline{D} \wedge \tag{6}$$

$$\overline{A} \cap B \subseteq \overline{C} \cap D \wedge \tag{7}$$

$$\overline{A} \cap \overline{B} \supseteq \overline{C} \cap \overline{D} \tag{8}$$

(5)–(7) are equivalent to Tversky's [10] monotonicity axiom and agree with his matching axiom (see page 3). (8) was added by Geist, Lengnink and Wille. It was intended to capture *context effects* on similarity: expanding the context (i.e. M) in which the objects are evaluated leads to an increase in similarity between objects, whose feature sets have not changed as a result of the expansion.

When (5)–(8) apply, $(A, B) \geq (C, D)$. '$\geq$' is a partial order. As usual, if $(A, B) \geq (C, D)$ and $(A, B) \leq (C, D)$, then $(A, B) = (C, D)$. If neither is true, then the pairs are incomparable: $(A, B) \neq (C, D)$. From (5)–(8) follows [3]:

$$(A, \overline{A}) \leq (X, Y) \leq (A, A) \Longleftrightarrow X = A \wedge$$
$$(\overline{B}, B) \leq (X, Y) \leq (B, B) \Longleftrightarrow Y = B \tag{9}$$

The idea stated in (9) that the attribute set most dissimilar to A is $\overline{A}$ and that the attribute set most similar to A is A itself seems very plausible.

3   Experimental Test of the GLW Model

The key prediction of the GLW model is incomparability of certain pairs of pairs of attribute sets, and thus incomparability of pairs of pairs of objects. This prediction sets it apart from other similarity models, which imply total ordering of pairs. We therefore hypothesized that participants, when given the choice, would choose this option as predicted by the GLW model. We conducted an experiment of two parts: in the first part, we asked participants to assign attributes to images of objects. In the main experiment, participants had to order pairs of objects by their perceived similarity, with an incomparability option. We then tried to predict their responses in the main experiment from the attribute assignment in the pre-test with the GLW model. All experimental procedures were approved by the local ethics commission of the department of Psychology at the University of Marburg, where the experiments were carried out. Participants were instructed in a standardized manner and gave written consent.

3.1 Feature List Experiment

The goal of our first experiment ($N = 22, 18♀$) was to collect the attribute set M for the objects (see left in Fig. 1) for the main experiment. The stimuli are modified versions of pictures taken from the website Flickr (https://www.flickr.com/) and were all released under a Creative Commons license. The experimental procedure was a modified version of Ahn and Dabbish's [1] ESP game. For a list of the creators of the images see [6]. Briefly, pairs of participants were presented 10 stimuli (see Fig. 1, left) in a randomized order. They were asked to type in either a noun or an adjective describing the presented stimulus. Subjects were specifically instructed to type in a *feature* of the object and not the name of the object. Their task was to think of features that the other participant would agree on. They advanced to the next stimulus after five agreements or five minutes, whichever came first.

The most frequent attributes for all stimuli, and the most frequent unique attribute for each stimulus are listed in Fig. 1, right. For a more detailed description of this part of the experiment, including the results, and the attribute selection process, see [6].

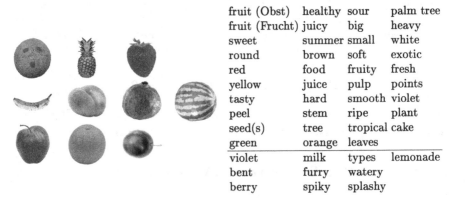

fruit (Obst)	healthy	sour	palm tree
fruit (Frucht)	juicy	big	heavy
sweet	summer	small	white
round	brown	soft	exotic
red	food	fruity	fresh
yellow	juice	pulp	points
tasty	hard	smooth	violet
peel	stem	ripe	plant
seed(s)	tree	tropical	cake
green	orange	leaves	
violet	milk	types	lemonade
bent	furry	watery	
berry	spiky	splashy	

Fig. 1. *Left*: stimuli that were used both in the pre-test and the main experiment. *Right*: above the line are the most frequent attributes for all stimuli. Below the line is each most frequent unique attribute for each stimulus. Words were translated from German. In one case two different German words translate to the same English one, hence the German one is given in parentheses.

3.2 Main Experiment

$N = 60, 40 ♀$ participants completed the study. They were either compensated with course credits or money (8€ per hour). The main experiment consisted of two parts: attribute selection and pair ordering. The same stimuli as in the feature list experiment were used in the main experiment, too. At the beginning of each of the two parts was a training block.

Procedures. During attribute selection participants were presented with one stimulus object at a time, and the attribute list (Fig. 1, right)[1]. They were asked to select all attributes that they saw as a "fitting description" (quote from the instruction) for the object. Stimuli were presented in randomized order.

During pair ordering participants were presented with two pairs of stimuli, one on the left, (a, b), and one on the right side of the screen, (c, d). They were asked to compare the similarity of the two pairs of stimuli (a, b), (c, d) and select one of the following four options: (1) (a, b) is more similar than (c, d), (2) (c, d) more similar than (a, b), (3) (a, b) is as similar as (c, d) and (4) (a, b) and (c, d) "cannot be compared with each other, meaning that [one] finds this comparison to be nonsensical" (quote from the instructions).

Each of these options represents an operationalisation of the four possible predictions of the GLW model: (1) corresponds to $(A, B) > (C, D)$, (2) to $(A, B) < (C, D)$, (3) to $(A, B) = (C, D)$ and (4) to $(A, B) \neq (C, D)$.

Since it was impractical to present every possible 4-tuple combination of 10 stimuli to each participant ($10^4 = 10000$) we could show only a selection of stimuli. While trying to strike a balance between a large enough selection to get a good idea of the participants' similarity judgement and avoiding to put too much strain on them, we decided to present 400 stimulus combinations, chosen such that the four predictions of the GLW model appeared in proportions that were as balanced as possible, excluding trivial cases. For further details on the selection of stimulus combinations, see [6].

Analysis. For the purpose of the data analysis we looked at the main experiment as a classification task, in which the GLW model was the classifier and the answers which the participants provided were considered as the ground truth data. The classes were the four possible answer options. We calculated the frequency of both the true and the predicted classes. We also calculated precision (fraction of participant's choices matching a given model prediction) and recall (fraction of model responses matching a given participant's choice) of the classifier.

Results. In Fig. 2, left, we picture the mean frequency ($\pm$ standard deviation) of the predictions of the GLW model and the answers across all participants. It is evident that there is a large mismatch between the human use and the GLW model's prediction of the different answer options.

---

[1] "violet" was only included once.

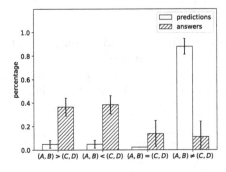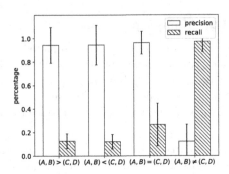

Fig. 2. Left: GLW model predictions of similarity pair orderings (open bars) and participants' answers (hatched bars). Shown are sample means ± on standard deviation. Incomparability is predicted too often. **Right:** precision (open bars) and recall (hatched bars) of the GLW model. Precision is high when pairs are comparable.

Given the small standard deviations, this mismatch is present across all participants. The main problem seems to be that the GLW model predicts incomparability far too often. Moreover, the frequency with which the GLW model makes this prediction is fairly constant across participants.

To elucidate the model's predictive power, we also computed the precision and the recall. The results are shown in Fig. 2, right. The precision for $<$, $>$ and $=$ is very high: whenever the model predicts comparability of any kind, this comparability is also perceived by the participants. This is a desirable feature of a model of human similarity perception. However, because it does so too infrequently, its recall for these answer options is rather low. This is in line with a very high recall and low precision for the $\neq$ option.

4   Discussion

The GLW model exhibits high precision ($>90\%$) whenever it predicts comparability. This is a desirable feature of a similarity model. However, with 87.9% the fraction of incomparability predictions is too high: the participants chose this option with a much smaller frequency of 11.2%.

Given that a non-zero fraction of answers were $\neq$, we believe that the cognitive ranking of similarities can follow a partial order, rather than a total order as predicted by other models or similarity indices derived thereof. Note that approximately one-fifth of the participants did not choose $\neq$ at all. It could be that the latter finding is simply an artifact of our experimental design. For example, it is conceivable that the instruction did not explain clearly enough what the answer $(A, B) \neq (C, D)$ means. Alternatively, this variability between participants might provide some evidence for the hypothesis that one major inter-individual difference in human similarity judgment could be partial versus total order.

The main question from our perspective is how to create a model that makes more comparability predictions in line with participants' similarity ratings. It seems that the GLW model is too strict: Whenever at least one of the four statements does not hold, GLW predicts incomparability. This can very easily be the case, for example when A, B, C and D each contain a unique attribute. In this case both (6) and (7) cannot hold. It might be useful to investigate models which drop some terms from the GLW model or replace conjunctions with disjunctions.

To conclude, we think that future models of perceptual similarity should preserve the GLW model's useful features: partial ordering of similarities and very high precision of the comparability predictions. But the overly large fraction of incomparability predictions needs to be remedied. We hope that our study inspires more research in this fascinating and practically relevant direction.

Acknowledgements. The authors were supported by the DFG-CRC-TRR 135 'Cardinal Mechanisms of Perception', project C6.

References

1. von Ahn, L., Dabbish, L.: Labeling images with a computer game. In: Proceedings of the SIGCHI Conference on Human Factors in Computing Systems. CHI 2004, pp. 319–326. ACM, New York (2004). https://doi.org/10.1145/985692.985733
2. Ganter, B., Wille, R.: Formal Concept Analysis Mathematical Foundations. Springer, Heidelberg (1999). https://doi.org/10.1007/978-3-642-59830-2
3. Geist, S., Lengnink, K., Wille, R.: An order-theoretic foundation for similarity measures. In: Lengnink, K. (ed.) Formalisierungen von Ähnlichkeit aus Sicht der Formalen Begriffsanalyse, pp. 75–87. Shaker Verlag (1996)
4. Kuznetsov, S.O.: On stability of a formal concept. Ann. Math. Artif. Intell. **49**(1–4), 101–115 (2007)
5. Murphy, G.L.: The Big Book of Concepts 1. MIT Press, Cambridge (2004). MIT Press paperback ed. A Bradford book
6. Schubert, M.: Empirische Überprüfung des generalisierten Kontrastmodells nach Lengnink. Geist und Wille, Marburg (2017)
7. Shepard, R.N.: The analysis of proximities: multidimensional scaling with an unknown distance function. I. Psychometrika **27**(2), 125–140 (1962)
8. Shepard, R.N.: The analysis of proximities: multidimensional scaling with an unknown distance function. II. Psychometrica **27**(3), 219–246 (1962)
9. Stumme, G., et al.: Conceptual clustering with iceberg concept lattices. In: Proceedings of GI-Fachgruppentreffen Maschinelles Lernen 2001. Universität Dortmund
10. Tversky, A.: Features of similarity. **84**(4), 327–352 (1977). https://doi.org/10.1037/0033-295X.84.4.327

Combining and Contrasting Formal Concept Analysis and APOS Theory

Uta Priss[✉][iD]

Zentrum für erfolgreiches Lehren und Lernen, Ostfalia University of Applied Sciences,
Wolfenbüttel, Germany
http://www.upriss.org.uk

Abstract. This paper investigates how two different theories (FCA and APOS Theory) complement each other with respect to applications in mathematics education research. APOS Theory is a constructivist theory concerned with mathematical learning whereas FCA is a mathematical theory itself. Together both theories provide different insights into how conceptual structures can be modelled and learned: FCA provides a model for a structural analysis of mathematical concepts and APOS Theory highlights the challenges involved in learning concepts that are complex and abstract.

1 Introduction

The question as to how a person learns mathematical concepts can be investigated from many different perspectives including cognitive and neuroscientific, mathematics educational, philosophical and structural considerations. For this paper we have selected two theories which provide very different views on what concepts are and how they may be learned, but which complement each other. The first theory, Formal Concept Analysis (FCA), was developed by Rudolf Wille in the 1980s as a mathematical lattice-based model of conceptual hierarchies with applications in data analysis and knowledge representation (Ganter and Wille 1999). The second theory, APOS Theory (e.g., Dubinsky and McDonald (2002)), was developed by Ed Dubinsky in the area of mathematics education research based on a constructivist understanding of learning. The abbreviation APOS stands for Action, Process, Object and Schema and is explained further below.

Some parallels can be observed between how FCA and APOS Theory were invented. Both founders (Wille and Dubinsky) started out as pure mathematicians at about the same time and then developed a deep interest in teaching and learning. Both were influenced by pedagogical, philosophical theories (Peirce and Piaget, respectively). Both theories have mathematical constructs at their core (lattices in FCA and functions in APOS Theory) which correspond to the mathematical research interests of their founders (general algebra and functional analysis). Both attracted sufficient interest to each establish a research community that is still active today.

© Springer International Publishing AG, part of Springer Nature 2018
P. Chapman et al. (Eds.): ICCS 2018, LNAI 10872, pp. 96–104, 2018.
https://doi.org/10.1007/978-3-319-91379-7_8

On the surface both theories are quite different. One theory (FCA) focuses on mathematical modelling which can be applied to educational data, but can also be used in many other domains. Its model of concepts is a mathematical abstraction of a philosophical understanding of what concepts are. Thus it is somewhat removed from cognitive and educational models of learning. The other theory (APOS Theory) is mainly concerned with what happens when students mentally construct concepts within an educational setting. Contrary to FCA, APOS Theory does not provide a formal, mathematical description of its notions. Nevertheless we argue in this paper that both theories are complementary to each other. Each provides viewpoints that are missing from the other theory, but might be beneficial to a more in-depth analysis of mathematical learning.

Section 2 briefly describes the two theories. It is followed, in Sects. 3 and 4, by a closer inspection of the transitions between the main stages suggested by APOS Theory and their relationship to FCA. The paper finishes with a conclusion.

2   A Short Description of the Two Theories

2.1   Formal Concept Analysis (FCA)

FCA is a theory of knowledge representation that presents a mathematical model for conceptual hierarchies using lattice theory (Ganter and Wille 1999). It formalises notions of classification, ordering, hierarchies and concepts. Because FCA has been a topic of this conference for many years, this paper does not include an introduction to FCA. This section argues that a notion of 'formal concept' provides a means for modelling concepts occurring in formal disciplines, such as mathematics, and explaining how these differ from concepts occurring in the mental processing of natural language.

The definition of 'concept' in FCA is a formal mathematical definition and does therefore not necessarily express exactly the same as what researchers from other disciplines perceive as concepts. Nevertheless the idea of a concept consisting of an extension and an intension is consistent with a philosophical notion of 'concept' that has "grown during centuries from Greek philosophy to late Scholastic and has finally found its modern formulation in the 17th century by the Logic of Port Royal" (Mineau et al. 1999, p. 432). Thus FCA formalises a pre-existing notion.

The duality of extension and intension occurs in many mathematical disciplines. Functions can be represented extensionally by n-tuples of values and intensionally by formulas. In set theory, sets can be defined either extensionally by listing their values or intensionally via a condition. Boolean logical statements (such as $a \vee a = a$) can be evaluated by either writing truth tables or by applying transformations and axioms. In some domains, this duality leads to interesting questions. For example the field of abstract algebraic logic combines extensional questions about algebras with intensional questions about the dimension and expressive power of axiom bases. Interestingly there are some algebras which are not representable which means that they can be described intensionally, but not extensionally. Other algebras (such as proper relation algebras) are extensionally easy to describe, but do not have a finite axiom basis (or intensional description).

From a cognitive viewpoint, Endres et al. (2010) show that FCA provides a model of neural representations of stimuli within the visual cortex of the brain. Thus it is possible that concepts in the sense of FCA are highly relevant for modelling actual brain activity. In other disciplines concepts are often perceived as fuzzy, context-dependent, embodied or prototypical structures. For example, a definition of a concept for 'democracy' does not have a universal, precise extension and intension. Priss (2002) calls such concepts 'associative' and observes that the duality of formal versus associative does not just apply to concepts but to many structures. Formal concepts tend to occur in mathematics and natural sciences. For example, while there is no universal formal definition of 'bird' in natural languages, the concept of 'passerine bird' is formally defined in biology and has a precise extension and intension at any point in time (which can change, but only if new scientifically relevant facts are discovered). This notion of 'formal concept' is slightly broader than the one used in FCA because the extension and intension of 'passerine bird' are precise and finite yet impossible to be exhaustively listed.

With respect to mathematics education, Priss (2018) argues that one reason for why many people find it difficult to learn mathematical concepts is because such concepts are strictly formal in nature, but learners think of them in an associative manner. This distinction is somewhat akin to Tall and Vinner's (1981) distinction between concept definition (formal) and concept image (associative). For example, the mathematical definition of 'graph' corresponds to a formal concept. But if one asks students who have just started to learn mathematics at university to define 'graph' they might state that it is something that is represented graphically. Thus they are describing an associative concept via a prototypical feature. But 'being graphically represented' is neither necessary nor sufficient for a graph and thus irrelevant for the formal concept of 'graph'.

The conclusion of this section is that if mathematical concepts are formal in nature and have precise extensions and intensions, then FCA concept lattices can be used to structure mathematical knowledge. The content of such concept lattices would be incomplete because it is not possible to list all elements in the extensions and intensions, and not everything is known (as mentioned in the example of abstract algebraic logic above). But contrary to other FCA applications where concept lattices present an interpretation of data, mathematical knowledge can be represented with FCA so that the conceptual hierarchy in the lattices corresponds to provable mathematical statements. For example, a concept for 'relation' would be a superconcept of 'function' in the lattice which can also be proven with mathematical theory. The relationships in the lattice then represent structures that need to be learned by a student of mathematics.

2.2 APOS Theory

Dubinsky's APOS Theory (Dubinsky and McDonald 2002) is based on Piaget's 'reflective abstraction' and states that mathematical knowledge is learned as a progression involving actions, processes, objects[1] and schemas. Actions consist of actually performing some transformation. For example, with respect to a function that produces the square of two numbers, an action-level understanding means that one can multiply a number with itself if shown how to do this. A process-level understanding of this function means that one can imagine or think about calculating squares without actually doing it. At this level one can calculate squares for actual numbers, but also for an unknown x or for complex expressions (for example $(a + b)^2$). Furthermore one understands that the function can be reversed (by obtaining a square root) and one can think about the behaviour of the function as it approaches infinity. Reaching a process-level understanding is thus a complex achievement. An object-level understanding of this function encapsulates the function itself into an object. This means that it can be composed with other functions and actions, processes and transformations can be applied to it. For example, a square function could be used as an input to another function.

A schema combines actions, processes and objects that belong together. Thus a schema of functions involves a general understanding of how they are used and how their actions, processes and objects relate to each other. Schemas can become objects themselves and can be combined with other schemas. But Arnon et al. (Arnon et al. 2013, p. 26) indicate that even within APOS Theory the notion of a schema is not completely finalised and more research about schemas is needed. A 'genetic decomposition' is a detailed description of a schema that shows dependencies between the mental constructions and thus can be used to determine the sequence in which the materials might be learned. Considering the conclusion of the last section about the possibilities of representing mathematical knowledge with FCA, the suggestion we are proposing in this paper is that FCA concept lattices could be helpful for representing the genetic decompositions of APOS Theory. Currently, genetic decompositions are designed by teachers based on experience and by using data from interviews with students. The representation of genetic decompositions is semi-formal and their relationships are a mixture of conceptual, part-whole and other relationships. Thus a genetic decomposition cannot be converted into a lattice. But we are proposing that lattices could be used as building blocks of genetic decompositions with other relationships possibly added to the lattices.

It should be emphasised that although the stages of APOS Theory indicate a progression, an individual does not always acquire these in this sequence. Sometimes a student might already know some process-level aspects while still being mostly in an action-level stage. Also a student can switch between and combine stages while working on a particular task. One important purpose of determining an APOS analysis of a particular mathematical topic is to devise a teaching and learning cycle using activities (which help students to make the required mental constructions), class discussions (during which a teacher observes if the students were successful at forming appropriate mental constructions and relates

---

[1] In FCA the notion 'object' traditionally has a different meaning. In order to avoid confusion, in this paper we use 'element' instead of 'object' for the FCA notion.

what has been learned to relevant mathematical knowledge) and exercises (which reinforce what has been learned and prepare students for the next iteration of the cycle). This is called the ACE learning cycle consisting of Activities, Class discussions and Exercises. The ACE learning cycle approach has been shown in numerous studies to be more effective than traditional lecture-centric teaching methods (Arnon et al. 2013).

This very brief description of APOS Theory is obviously incomplete (cf. Arnon et al. (2013) for a comprehensive discussion of APOS Theory). It is possible that APOS Theory is mainly only relevant for the learning of mathematics and similar formal domains. An example from music education below shows that the transition from action to process (called 'interiorisation') can occur in other domains. But the transition from process to object (called 'encapsulation') involves some form of abstraction and may only be relevant for domains where abstraction is frequently encountered. The next two sections further investigate interiorisation and encapsulation, respectively.

3   Interiorisation and Conceptualisation

All core notions of APOS Theory are constructivist and focus on mental constructions. Nevertheless it is argued in this paper that it might be of interest to compare them to similar notions from other disciplines. For example, because interiorisation implies that internal structures are created and because, as mentioned above, experts can deliberately switch between different APOS stages, it should be remarked that it is also possible to consciously externalise some internal thought. Scaife and Rogers (1996) coined the notion of 'external cognition' in the context of understanding how graphical representations work. It refers to using external representations during cognitive processes. For example, mathematical tasks may become easier when they are conducted with pen and paper. External cognition implies that external representations may add a significant cognitive value by making something easier which would be difficult otherwise.

We propose considering interiorisation to be an example of conceptualisation because understanding a mathematical concept involves understanding which other concepts are equivalent or implied by the concept and which statements can be proven about the concept (including which elements are in its extension and which attributes in its intension). We argue that processes, objects and schemas correspond to formal mathematical concepts. Actions are not concepts because they are scripted procedures which do not require any understanding.

A further aspect of interiorisation is often the creation of a function. Dubinsky considers processes usually as functions and being able to write something as a function is a good indicator that a student has achieved a process-level understanding (cf. Arnon et al. 2013, p. 199) even though "a Process is only one part of a function". Not every concept corresponds to a function, but in mathematics many concepts can be represented by functions. With respect to the fact that concepts in FCA consist of extensions and intensions this suggests that it might be of interest in some applications to represent intensions of FCA concepts by functions (e.g. algorithms or procedures) instead or in addition to lists of attributes.

An interesting question is whether APOS Theory is also relevant for other disciplines. We argue that while encapsulation is only relevant if the domain is sufficiently abstract, interiorisation might be observed in other domains. For example, in music education, one can observe that music learners at first often do not perceive rests at all. At an action-level understanding they might observe a rest by making a conscious effort to be silent for the required amount of time. A process-level understanding of musical rests involves perceiving rests as an integral part of music where errors with respect to rests produce the same sense of incorrectness as an imprecise pitch or a false articulation. Interiorisation then corresponds to a cognitive change of perception or conceptualisation from something that is consciously performed to something that is internally felt. In this case the resulting process (or concept) is neither a mathematical function nor a formal concept.

4   Encapsulation and Switching to a Meta-Level

The process/object transition (or encapsulation) has been extracted from APOS Theory and been used in other theories, such as by Hazzan (2003). As stated above, we believe that encapsulation always involves some form of abstraction. In FCA terminology it corresponds to a concept from one formal context becoming an element in another formal context and thus requiring a meta-level. In mathematics this happens frequently and is possibly unlimited. Mathematical category theory provides an example where even the basic elements are already encapsulated. An example from another discipline is the notion of 'emic units' in linguistics which arose from the observation that the distinction between certain units (such as phoneme and phone) also applies to other linguistic units (morpheme, grapheme, lexeme etc) and thus presents a general (meta-level) structure of semiotic systems. But the number of levels in non-mathematical domains is limited.

We argue in this section that encapsulation has cognitive, structural and systems-theoretical aspects. An example of cognitive aspects of encapsulation is chunking. Cognitive scientists consider grouping information so that it becomes easier to process and memorise an innate feature of human cognition. For example, Miller (1956) observes that humans can hold only about 7 items in short-term memory. Thus if one wants to memorise something that contains more than 7 items, it needs to be subdivided into groups of not more than 7 items each. Experiments have, for example, shown that expert chess players are able to memorise the configuration of an entire chess board because they chunk it into subgroups (Gobet et al. 2001). It is not necessary for the subgroups to form a unit that is meaningful apart from aiding as a memorisation task. Thus not every chunk becomes a meta-level object. But the human brain seems to have a tendency to form chunks or gestalts even if they are not really meaningful, such as seeing pictures in the clouds.

Part-whole relationships are an example of structural aspects of encapsulation. Linguistic analyses have shown that meronymic, or part-whole relationships, are core structures of language (Miller et al. 1990) and thus of human thinking. A large number of different types of linguistic part-whole relations have been identified in the literature (cf. Priss (1996) for some examples). In philosophy there is an entire discipline (mereology) dedicated to a theory of part-whole relationships. Encapsulation creates objects which are wholes consisting of parts, but not every part-whole relationship creates a meta-level. For example, the concepts of 'finger', 'hand' and 'limb' are all at the same level of abstraction even though a finger is part of a hand which is part of a body. But the concept 'body parts' is at a meta-level and has 'finger', 'hand' and 'limb' in its extension.

A meta-level object has parts, but the parts are not necessarily relevant to how the object behaves itself. This relates to emergence which is a systems-theoretical aspect of encapsulation. In cognitive science and related fields the notion of 'emergence' is used to describe objects or features that emerge from interactions among multiple elements in a system without having a simple relationship between the original elements and the emerging objects (Clark 1997). There are many examples of emergent phenomena, such as flight patterns among birds, weather phenomena, crystalline structures and physical properties that occur at macroscopic scales, but not at microscopic scales. Often the initial elements follow simple rules. For example, migrating birds arrange their position with respect to the other flying birds using simple rules. In John Conway's game of life[2] there are a few simple rules about cell behaviour which cause complex patterns to emerge. In these cases, there is a causal connection between the original system with its elements and the emerging objects, but it is not possible to describe this connection using a simple input/output mapping (Clark 1997).

A claim of this paper is that chunking and part-whole relationships are necessary, but not sufficient features of encapsulation. Emergence is often a consequence of encapsulation. Any mathematical definition of an object with some operations or properties gives rise to other objects and properties. The purpose of encapsulation is so that the resulting object can be used in further processes and transformations. By becoming an independent unit and interacting with other objects and processes, new features and properties can be observed which were not purposefully created during the encapsulation and which have nothing to do with the properties of the original processes. For example, even though multiplication of numbers is commutative, matrix multiplication is not. There is no intuitive transfer of properties from parts to whole.

5   Conclusion

This paper argues that FCA and APOS Theory complement each other with respect to analysing how mathematical concepts are learned. FCA provides a model for mathematical knowledge as concept lattices of formal concepts. This

---

[2] http://en.wikipedia.org/wiki/Conway's_Game_of_Life.

clarifies for example differences between part-whole and subconcept relationships and the transition to a meta-level when a concept from one context becomes an element in another formal context. Concept lattices can also explain why some concepts are more difficult to be learned than others: If a formal context is expanded by adding further new elements and attributes, it can happen that the new lattice is quite similar to the previous one if the new elements and attributes do not 'disturb' the previous connections. But, for example, if a new attribute applies to many old elements which did not have anything in common in the previous lattice, then the new lattice might be radically different. Other interesting research questions for FCA are whether it might be useful in some applications to represent intensions by functions and how the relationship between a lattice and a meta-level lattice which has as elements the concepts of the other lattice can be characterised.

APOS Theory emphasises the difficulties related to learning mathematics because of challenging conceptualisations (interiorisations), encapsulations which lead from one level of abstraction to a meta-level and the complexity of schemas that encompass mathematical knowledge. FCA could potentially be employed[3] to derive a more formal representation of genetic decompositions which represent schemas and structure the sequence in which mathematical concepts can be learned. Thus FCA and APOS Theory complement each other, and a combination of both might provide insights about mathematical learning which surpass the analytic capabilities of either theory by itself.

References

Arnon, I., Cottrill, J., Dubinsky, E., Oktac, A., Roa Fuentes, S., Trigueros, M., Weller, K.: APOS Theory - A Framework for Research and Curriculum Development in Mathematics Education. Springer, Heidelberg (2013). https://doi.org/10.1007/978-1-4614-7966-6

Clark, A.: Being There. Putting Brain, Body, and World Together Again. MIT Press, Cambridge (1997)

Dubinsky, E., Mcdonald, M.A.: APOS: a constructivist theory of learning in undergraduate mathematics education research. In: Holton, D., Artigue, M., Kirchgräber, U., Hillel, J., Niss, M., Schoenfeld, A. (eds.) The Teaching and Learning of Mathematics at University Level. NISS, vol. 7, pp. 275–282. Springer, Dordrecht (2001). https://doi.org/10.1007/0-306-47231-7_25

Endres, D.M., Foldiak, P., Priss, U.: An application of formal concept analysis to semantic neural decoding. Ann. Math. Artif. Intell. **57**(3), 233–248 (2010). Springer

Ganter, B., Wille, R.: Formal Concept Analysis. Mathematical Foundations. Springer, Heidelberg (1999). https://doi.org/10.1007/978-3-642-59830-2

Gobet, F., Lane, P.C.R., Croker, S., Cheng, P.C.-H., Jones, G., Oliver, I., Pine, J.: Chunking mechanisms in human learning. Trends Cogn. Sci. **5**(6), 236–243 (2001)

Hazzan, O.: How students attempt to reduce abstraction in the learning of mathematics and in the learning of computer science. Comput. Sci. Educ. **13**(2), 95–122 (2003)

---

[3] Possibly in combination with conceptual graphs (Sowa 2008) and concept graphs (Wille 2002).

Miller, G.A.: The magical number seven, plus or minus two: some limits on our capacity for processing information. Psychol. Rev. **63**(2), 81–97 (1956)

Miller, G.A., Beckwith, R., Fellbaum, C., Gross, D., Miller, K.: Introduction to Word-Net: an on-line lexical database. Int. J. Lexicogr. **3**(4), 235–244 (1990)

Mineau, G., Stumme, G., Wille, R.: Conceptual structures represented by conceptual graphs and formal concept analysis. In: Tepfenhart, W.M., Cyre, W. (eds.) ICCS-ConceptStruct 1999. LNCS (LNAI), vol. 1640, pp. 423–441. Springer, Heidelberg (1999). https://doi.org/10.1007/3-540-48659-3_27

Priss, U.: Classification of meronymy by methods of relational concept analysis. In: Proceedings of the 1996 Midwest Artificial Intelligence Conference, Bloomington, Indiana (1996)

Priss, U.: Associative and formal structures in AI. In: Proceedings of the 13th Midwest Artificial Intelligence and Cognitive Science Conference, Chicago, pp. 36–42 (2002)

Priss, U.: A semiotic-conceptual analysis of conceptual development in learning mathematics. In: Presmeg, N., Radford, L., Roth, W.-M., Kadunz, G. (eds.) Signs of Signification. IM, pp. 173–188. Springer, Cham (2018). https://doi.org/10.1007/978-3-319-70287-2_10

Scaife, M., Rogers, Y.: External cognition: how do graphical representations work. Int. J. Hum.-Comput. Stud. **45**, 185–213 (1996)

Sowa, J.: Conceptual Graphs. Foundations of Artificial Intelligence, pp. 213–237. Elsevier, New York (2008)

Tall, D., Vinner, S.: Concept image and concept definition in mathematics with particular reference to limits and continuity. Educ. Stud. Math. **12**(2), 151–169 (1981)

Wille, R.: Existential concept graphs of power context families. In: Priss, U., Corbett, D., Angelova, G. (eds.) ICCS-ConceptStruct 2002. LNCS (LNAI), vol. 2393, pp. 382–395. Springer, Heidelberg (2002). https://doi.org/10.1007/3-540-45483-7_29

Musical Descriptions Based on Formal Concept Analysis and Mathematical Morphology

Carlos Agon[1], Moreno Andreatta[1,2(✉)], Jamal Atif[3], Isabelle Bloch[4], and Pierre Mascarade[3]

[1] CNRS-IRCAM-Sorbonne Université, Paris, France
{carlos.agon,moreno.andreatta}@ircam.fr
[2] IRMA/GREAM/USIAS, Université de Strasbourg, Paris, France
[3] Université Paris-Dauphine, PSL Research University, CNRS, UMR 7243, LAMSADE, 75016 Paris, France
jamal.atif@dauphine.fr, pierre.m@protonmail.com
[4] LTCI, Télécom ParisTech, Université Paris-Saclay, Paris, France
isabelle.bloch@telecom-paristech.fr

Abstract. In the context of mathematical and computational representations of musical structures, we propose algebraic models for formalizing and understanding the harmonic forms underlying musical compositions. These models make use of ideas and notions belonging to two algebraic approaches: Formal Concept Analysis (FCA) and Mathematical Morphology (MM). Concept lattices are built from interval structures whereas mathematical morphology operators are subsequently defined upon them. Special equivalence relations preserving the ordering structure of the lattice are introduced in order to define musically relevant quotient lattices modulo congruences. We show that the derived descriptors are well adapted for music analysis by taking as a case study Ligeti's String Quartet No. 2.

Keywords: Computational music analysis · Formal concept analysis
Mathematical morphology · Congruences · Quotient lattices
Harmonico-morphological descriptors · Musical information research

1 Introduction

Despite a long historical relationship between mathematics and music, computational music analysis is a relatively recent research field. In contrast to statistical methods and signal-based approaches currently employed in Music Information Research (or MIR[1]), the paper at hand stresses the necessity of introducing

---

[1] Following the Roadmap described in [18], we prefer to consider MIR as the field of Music Information Research instead of limiting the scope to purely Music Information Retrieval. This approach constitutes the core of an ongoing research project entitled SMIR (Structural Music Information Research: Introducing Algebra, Topology and Category Theory into Computational Musicology). See http://repmus.ircam.fr/moreno/smir.

© Springer International Publishing AG, part of Springer Nature 2018
P. Chapman et al. (Eds.): ICCS 2018, LNAI 10872, pp. 105–119, 2018.
https://doi.org/10.1007/978-3-319-91379-7_9

a structural multidisciplinary approach into computational musicology making use of advanced mathematics. It is based on the interplay between algebra and topology, and opens promising perspectives on important prevailing challenges, such as the formalization and representation of musical structures and processes, or the automatic classification of musical styles. It also differs from traditional applications of mathematics to music in aiming to build bridges between different musical genres, ranging from contemporary art music to popular music, and therefore showing the generality of the conceptual tools introduced into computational musicology. Most of these tools belong to the domain of Formal Concept Analysis (FCA), a research field that has been introduced in the beginning of the 1980s by Rudolf Wille and, independently, by Marc Barbut and Louis Frey [3], in both cases as an attempt at reconstructing Lattice Theory [20, 23].[2] Interestingly, music was a major inspirational field for applying formal concept analysis, within the Darmstadt tradition. In his introductory essay on the link between mathematics and music [22], R. Wille proposed to represent the chords of the diatonic scale as equivalence classes, leading to a chord concept lattice. Following this seminal work, a morphology of chords has been proposed by Noll [13], showing the interest of this approach for computational music analysis. Some recent work renewed the interest of this approach as an original way to represent musical structures within the field of Music Information Research, by stressing the underlying algebraic aspects, with application of ordered structures and concept lattices to the algebraic enumeration and classification of musical structures for computational music analysis [16, 17].[3]

In this paper, we propose a way to combine algebraic formalizations and lattice-based representations of harmonic structures within existing musical compositions. Instead of analyzing the musical pitch content,[4] formal concept lattices are built from intervallic structures. The objective is to provide a way of summarizing the harmonic musical content by studying the properties of the underlying lattice organization. We make use of operators belonging to Mathematical Morphology (MM) which are defined on such lattices. This enables to define congruences between elements of the lattice and associated quotient lattices. These quotient lattices are in fact the symbolic and structural descriptors of the musical pieces that we propose to use as a generic tool for computational music analysis. As a case study, we show how the first movement of Ligeti's String Quartet No. 2 can be compactly described with three of such quotient lattices.

This paper is organized as follows. In Sect. 2 we summarize previous work on the definition of lattices of harmonic structures by means of their interval

---

[2] See [21] for an interesting discussion on the mutual influences between the Darmstadt school on Formal Concept Analysis and the French tradition on *Treillis de Galois*.

[3] See the MUTABOR language (http://www.math.tu-dresden.de/~mutabor/) for a music programming language making use of the FCA-based Standard Language for Music Theory [12] originally conceived by Rudolf Wille and currently developed at the University of Dresden.

[4] Note that, at this stage, the time information is not taken into account, and a musical excerpt is considered as an unordered set of chords.

content. In Sect. 3 we recall some definitions of mathematical morphology on complete lattices and propose specific operators (dilation and erosion) on musical concept lattices. The main original contribution of this paper is contained in Sect. 4 where we define a way to reduce a given concept lattice to its "core" structure via congruence relations. The resulting quotient lattices are precisely the structural descriptors used in the representation of a given musical piece. This opens new challenging perspectives for automatic music analysis and structural comparison between musical pieces of different styles.

2   Lattice of Interval Structures

In this section we recall how a concept lattice can be built from harmonic forms (as objects) and intervals (as attributes) [16,17].

Definition 1 (Harmonic system). *Let T be a set, $I = (I, +, -, 0)$ an Abelian group, and $\Delta : T \times T \to I$ a mapping such that $\forall t_1, t_2, t_3 \in T$:*

$$\Delta(t_1, t_2) + \Delta(t_2, t_3) = \Delta(t_1, t_3) \quad and \quad \Delta(t_1, t_2) = 0 \quad iff \quad t_1 = t_2.$$

Then the triplet $\mathbb{T} = (T, \Delta, I)$ is called algebraic harmonic system. Elements of T are tones and any subset of T is a chord. Elements of I are musical intervals.

Here we consider $\mathbb{T}_n = (\mathbb{Z}_n, \Delta_n, \mathbb{Z}_n)$, where $n \in \mathbb{Z}_+$ represents an octave, $\mathbb{Z}_n = \mathbb{Z}/n\mathbb{Z}$, and Δ_n is the difference modulo n. All chords are then projected in $\mathbb{T}_n$ using a canonical homomorphism. Moreover, two chords having the same number of notes (or chromas) and the same intervals between notes (i.e. defined up to a transposition) are considered equivalent, thus defining harmonic forms.

Definition 2 (Harmonic forms). *The set $\mathcal{H}(\mathbb{T}_n)$ of the harmonic forms of $\mathbb{T}_n$ is composed by the equivalence classes of the following equivalence relation Ψ:*

$$\forall H_1 \subseteq \mathbb{Z}_n, \forall H_2 \subseteq \mathbb{Z}_n, \ H_1 \Psi H_2 \ iff \ \exists i \ s.t. \ H_1 = H_2 + i$$

where $H + i = \{t + i \mid t \in H\}$ if $t + i$ exists for all $t \in H$.

In the sequel, we will use the following notation: $I_{H,t} = \{\Delta(t, t') \mid t' \in H\}$.

Definition 3 (Musical formal context). *A musical formal context, denoted by $\mathbb{K} = (\mathcal{H}(\mathbb{T}_n), \mathbb{Z}_n, R)$ is defined by considering harmonic forms, in $G = \mathcal{H}(\mathbb{T}_n)$, as objects and intervals, in $M = \mathbb{Z}_n$, as attributes. The relation R is defined from the occurrence of an interval in an harmonic form. A formal concept is a pair (X, Y), $X \subseteq G, Y \subseteq M$ such that $X \times Y \subseteq R$ and that is maximal for this property. The concept lattice $(\mathbb{C}(\mathbb{K}), \preceq)$ is then defined from the formal context and the partial ordering $\preceq$ defined as:*

$$(X_1, Y_1) \preceq (X_2, Y_2) \Leftrightarrow X_1 \subseteq X_2 (\Leftrightarrow Y_2 \subseteq Y_1).$$

For $X \subseteq G$ and $Y \subseteq M$, the derivation operators α and β are defined as $\alpha(X) = \{m \in M \mid \forall g \in X, (g, m) \in R\}$, and $\beta(Y) = \{g \in G \mid \forall m \in Y, (g, m) \in R\}$. The pair (α, β) induces a Galois connection between the partially ordered power sets $(\mathcal{P}(G), \subseteq)$ and $(\mathcal{P}(M), \subseteq)$, i.e. $X \subseteq \alpha(Y)$ iff $Y \subseteq \beta(X)$. Then the pair (X, Y) (with $X \subseteq G$ and $Y \subseteq M$) is a formal concept if and only if $\alpha(X) = Y$ and $\beta(Y) = X$ (X is then called extent and Y intent of the formal concept).

As in any concept lattice, the supremum and infimum of a family of concepts $(X_t, Y_t)_{t \in T}$ are:

$$\wedge_{t \in T}(X_t, Y_t) = \left(\cap_{t \in T} X_t, \alpha\big(\beta(\cup_{t \in T} Y_t)\big)\right), \tag{1}$$

$$\vee_{t \in T}(X_t, Y_t) = \left(\beta\big(\alpha(\cup_{t \in T} X_t)\big), \cap_{t \in T} Y_t\right), \tag{2}$$

Example 1. As a running example in this paper, we consider 7-tet $\mathbb{T}_7$ (e.g. the diatonic scale C, D, E, F, G, A, B). Let us define the formal context $\mathbb{K} = (\mathcal{H}(\mathbb{T}_7), \mathbb{Z}_7, R)$, where R is a binary relation such that for any harmonic form $F \in \mathcal{H}(\mathbb{T}_7)$ and any interval $i \in \mathbb{Z}_7$, we have $(F, i) \in R$ iff there exists $t \in F$ such that $i \in I_{F,t}$. Intervals are denoted by the index of the last note from the starting one, hence for the 7-tet intervals are unison (0), second (1), third (2), fourth (3). Note that other intervals (4, 5, and 6) are derived from these basic ones by group operations (inversion modulo octave).

Figure 1 illustrates the formal context $\mathbb{K} = (\mathcal{H}(\mathbb{T}_7), \mathbb{Z}_7, R)$ and the concept lattice $(\mathbb{C}(\mathbb{K}), \preceq)$ as defined in Definition 3.

This representation can be further enriched by considering interval multiplicities in harmonic forms [17] (for instance $\{0, 1\}$ contains two unisons and one second). Such representations will be used in Sect. 4.

3 Mathematical Morphology Operations on Musical Concept Lattices

3.1 Preliminaries

Let us recall the algebraic framework of mathematical morphology. Let $(\mathcal{L}, \preceq)$ and $(\mathcal{L}', \preceq')$ be two complete lattices (which do not need to be equal). All the following definitions and results are common to the general algebraic framework of mathematical morphology in complete lattices [4,5,7,8,11,15,19]. Note that different terminologies can be found in different lattice theory related contexts (refer to [14] for equivalence tables).

Definition 4. *An operator $\delta \colon \mathcal{L} \to \mathcal{L}'$ is an algebraic dilation if it commutes with the supremum (sup-preserving mapping):*

$$\forall(x_i) \in \mathcal{L}, \ \delta(\vee_i x_i) = \vee'_i \delta(x_i),$$

where $\vee$ (respectively $\vee'$) denotes the supremum associated with $\preceq$ (respectively $\preceq'$).

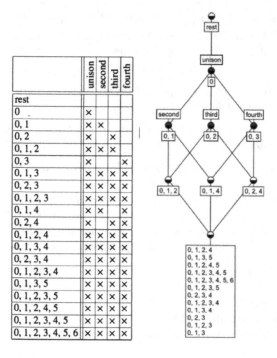

	unison	second	third	fourth
rest				
0	×			
0, 1	×	×		
0, 2	×		×	
0, 1, 2	×	×	×	
0, 3	×			×
0, 1, 3	×	×	×	×
0, 2, 3	×	×	×	×
0, 1, 2, 3	×	×	×	×
0, 1, 4	×	×		×
0, 2, 4	×		×	×
0, 1, 2, 4	×	×	×	×
0, 1, 3, 4	×	×	×	×
0, 2, 3, 4	×	×	×	×
0, 1, 2, 3, 4	×	×	×	×
0, 1, 3, 5	×	×	×	×
0, 1, 2, 3, 5	×	×	×	×
0, 1, 2, 4, 5	×	×	×	×
0, 1, 2, 3, 4, 5	×	×	×	×
0, 1, 2, 3, 4, 5, 6	×	×	×	×

Fig. 1. Formal context $\mathbb{K} = (\mathcal{H}(\mathbb{T}_7), \mathbb{Z}_7, R)$ and concept lattice $(\mathbb{C}(\mathbb{K}), \preceq)$ (reproduced from [17]).

An operator $\varepsilon \colon \mathcal{L}' \to \mathcal{L}$ is an algebraic erosion *if it commutes with the infimum (inf-preserving mapping)*:

$$\forall (x_i) \in \mathcal{L}', \ \varepsilon(\wedge'_i x_i) = \wedge_i \varepsilon(x_i),$$

where $\wedge$ and $\wedge'$ denote the infimum associated with $\preceq$ and $\preceq'$, respectively.

This general definition allows defining mathematical morphology operators such as dilations and erosions in many types of settings, such as sets, functions, fuzzy sets, rough sets, graphs, hypergraphs, various logics, etc., based on their corresponding lattices.

Algebraic dilations δ and erosions ε are increasing operators; moreover δ preserves the smallest element and ε preserves the largest element.

A fundamental notion in this algebraic framework is the one of adjunction.

Definition 5. *A pair of operators (ε, δ), $\delta \colon \mathcal{L} \to \mathcal{L}'$, $\varepsilon \colon \mathcal{L}' \to \mathcal{L}$, defines an* adjunction *if*

$$\forall x \in \mathcal{L}, \forall y \in \mathcal{L}', \delta(x) \preceq' y \iff x \preceq \varepsilon(y).$$

Note that the notion of adjunction corresponds to the Galois connection by reversing the order of either $\mathcal{L}$ or $\mathcal{L}'$. This induces a first direct link between derivation operators α, β on the one hand, and δ, ε on the other hand. Further links between FCA and MM have been investigated in [1,2].

Some important properties, that will be used in the following, are summarized as follows.

Proposition 1 *(e.g. [8,15]). If a pair of operators (ε, δ) defines an adjunction, then the following results hold:*

- *δ preserves the smallest element and ε preserves the largest element;*
- *δ is a dilation and ε is an erosion (in the sense of Definition 4).*

Let δ and ε be two increasing operators such that $\delta\varepsilon$ is anti-extensive and $\varepsilon\delta$ is extensive. Then (ε, δ) is an adjunction.

The following representation result also holds. If ε is an increasing operator, it is an algebraic erosion if and only if there exists δ such that (ε, δ) is an adjunction. The operator δ is then an algebraic dilation and can be expressed as $\delta(x) = \wedge'\{y \in \mathcal{L}' \mid x \preceq \varepsilon(y)\}$. A similar representation result holds for erosion.

All these results hold in the particular case of a concept lattice.

Particular forms of dilations and erosions can be defined based on the notion of structuring element, which can be a neighborhood relation or any binary relation [5,19]. In particular, such structuring elements can be defined as the balls of a given distance. This has been investigated in concept lattices, using several distances, in [1,2].

In the next sections we describe two examples of dilations and erosions, defined on the lattice $\mathbb{C}$, used to handle musical format contexts. They have been implemented in SageMath.[5] The other definitions proposed in [1,2] could be exploited as well for musical concept lattices.

3.2 Dilations and Erosions from the Decomposition into Join or Meet Irreducible Elements

The first example relies on the decomposition of a concept in a join-irreducible form (hence suitable for defining dilations), respectively meet-irreducible for defining erosions.

Definition 6 (Join and meet irreducible element). *An element a of a lattice $\mathbb{C}$ is join (respectively meet) irreducible if it is not equal to the least element of the lattice (respectively the largest element) and $\forall(a, b) \in \mathbb{C}^2, a = b \vee c \Rightarrow a = b$ or $a = c$ (respectively $a = b \wedge c \Rightarrow a = b$ or $a = c$).*

Any element of the lattice can be written (usually not uniquely) as the join (respectively meet) of some irreducible elements.

Since a dilation (respectively erosion) is defined as an operator that commutes with the supremum (respectively infimum), it is sufficient to define these operators on join (respectively meet) irreducible elements to extend them to any element of the lattice. This will be exploited next, in the proposed algorithm in Sect. 4.

---
[5] http://www.sagemath.org/.

3.3 Dilations and Erosions Based on Structuring Elements Derived from a Valuation

In the second example, we define dilations and erosions, based on structuring elements that are balls of a distance derived from a valuation on the lattice. In the following we propose to use valuations defined as the cardinality of filters or ideals.

Definition 7 (Filter and ideal). *Let a be an element of a lattice $\mathbb{C}$. The filter and ideal associated with a are the subsets of $\mathbb{C}$ defined as:*

$$F(a) = \{b \in \mathbb{C} \mid a \preceq b\}$$

$$I(a) = \{b \in \mathbb{C} \mid b \preceq a\}$$

Definition 8. *Let $(\mathbb{C}, \preceq)$ be a concept lattice. A real-valued function w on $(\mathbb{C}, \preceq)$ is a* lower valuation *if it satisfies the following (supermodular) property:*

$$\forall (a_1, a_2) \in \mathbb{C}^2, w(a_1) + w(a_2) \leq w(a_1 \wedge a_2) + w(a_1 \vee a_2), \tag{3}$$

and is an upper valuation *if it satisfies the following (submodular) property:*

$$\forall (a_1, a_2) \in \mathbb{C}^2, w(a_1) + w(a_2) \geq w(a_1 \wedge a_2) + w(a_1 \vee a_2) \tag{4}$$

A real-valued function is increasing (isotone) if $a_1 \preceq a_2$ implies $w(a_1) \leq w(a_2)$ and decreasing (antitone) if $a_1 \preceq a_2$ implies $w(a_1) \geq w(a_2)$.

Proposition 2 ([9,10]). *Let w be a real-valued function on a concept lattice $(\mathbb{C}, \preceq)$. Then the function defined as:*

$$\forall (a_1, a_2) \in \mathbb{C}^2, \ d_w(a_1, a_2) = 2w(a_1 \wedge a_2) - w(a_1) - w(a_2) \tag{5}$$

is a pseudo-metric if and only if w is a decreasing upper valuation.
The function defined as:

$$\forall (a_1, a_2) \in \mathbb{C}^2, \ d_w(a_1, a_2) = w(a_1) + w(a_2) - 2w(a_1 \vee a_2) \tag{6}$$

is a pseudo-metric if and only if w is a decreasing lower valuation.

Proposition 3 (Valuation from a filter or ideal). *Let w_F be the mapping defined on a concept lattice $\mathbb{C}$ as $\forall a \in \mathbb{C}, w_F(a) = |F(a)|$ where F is the filter associated with a. Then w_F is a decreasing lower valuation, i.e.*

$$\forall (a_1, a_2) \in \mathbb{C}^2, w_F(a_1) + w_F(a_2) \leq w_F(a_1 \wedge a_2) + w_F(a_1 \vee a_2)$$

The mapping d from $\mathbb{C} \times \mathbb{C}$ into $\mathbb{R}^+$ defined as $\forall (a_1, a_2) \in \mathbb{C}^2, d(a_1, a_2) = w_F(a_1) + w_F(a_2) - 2w_F(a_1 \vee a_2)$ is therefore a pseudo-distance.
Similarly, a pseudo-distance can be defined from the cardinality of the ideals.

Once the distance is defined, a structuring element is defined as a ball of this distance, for a given radius n. Dilations and erosions can then be written as:

$$\forall A \subseteq \mathbb{C}, \delta(A) = \{b \in \mathbb{C} \mid d(b, A) \leq n\}$$

$$\forall A \subseteq \mathbb{C}, \varepsilon(A) = \{b \in \mathbb{C} \mid d(b, \mathbb{C} \setminus A) > n\} = \mathbb{C} \setminus \delta(\mathbb{C} \setminus A)$$

where $d(b, A) = \min_{a \in A} d(b, a)$. Note that such dilations and erosions can also be applied to irreducible elements, in order to derive dilations and erosions using the commutativity with the supremum or infimum, as will be used in Sect. 4.

4   Harmonico-Morphological Descriptors Based on Congruence Relations

By using the concepts we have previously introduced, we define a way of reducing a concept lattice via some equivalence relations, namely congruences. This ideas goes back to Birkhoff [4], and was used in several works such as [6] with extension to non transitive relations (tolerance relations). Here we propose new congruences based on mathematical morphology.

4.1   Definitions

Definition 9 (Congruences and quotient lattices). *An equivalence relation θ on a lattice $\mathcal{L}$ is a congruence if it is compatible with join and meet, i.e. $(\theta(a, b)$ and $\theta(c, d)) \Rightarrow (\theta(a \vee c, b \vee d)$ and $\theta(a \wedge c, b \wedge d))$, for all $a, b, c, d \in \mathcal{L}$. This equivalence relation allows defining a quotient lattice which will be denoted as $\mathcal{L}/\theta$.*

Hence $\vee$ and $\wedge$ induce joint and meet operators on the quotient lattice $\mathcal{L}/\theta$. By denoting $[a]_\theta$ the equivalence class of $a \in \mathcal{L}$ for θ, we have $[a]_\theta \vee [b]_\theta = [a \vee b]_\theta$, and a similar relation for $\wedge$ (note that the same notations are used on $\mathcal{L}$ and on $\mathcal{L}/\theta$ for meet and join, when no ambiguity occurs). This way of defining quotient lattices enables to transfer the structure from the original concept lattice to the reduced one, therefore preserving the order relations between the elements.

Example 2. Let us apply these notions to the concept lattice $(\mathbb{C}(\mathbb{K}), \preceq)$ associated with the musical formal concept defined from the 7-tet, as before (see Example 1), using the interval multiplicities. In this diatonic space, one may define a formal equivalence relation between major/minor chords and major/minor seventh chords allowing us to reduce the initial lattice to the corresponding quotient lattice. More precisely, the congruence relation is defined in order to group $\{0, 2, 4\}$ and $\{0, 1, 3, 5\}$ into the same class. The other classes are derived so as to preserve the ordering relations between concepts. This reduction process is represented in Fig. 2.[6] For instance, let us consider $\{0, 1\}$ (we only mention the extent of the concepts here). We have $\{0, 2, 4\} \wedge \{0, 1\} = \{0\}$ and

---

[6] Note that the lattice contains the nodes representing the harmonic forms, and additional nodes, labeled with an arbitrary number, arising from the completion to obtain a complete lattice [16]. These intermediate nodes are already present in Wille's original lattice-based formalization of the diatonic scale.

$\{0, 1, 3, 5\} \wedge \{0, 1\} = \{0, 1\}$, which is consistent with the fact that $\{0\}$ and $\{0, 1\}$ are in the same congruence class.

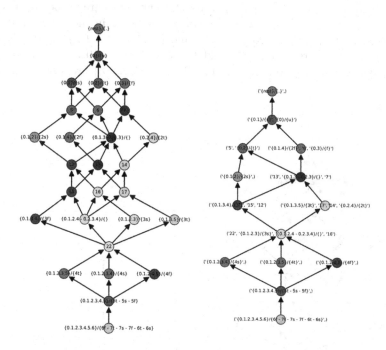

Fig. 2. Left: concept lattice, where all concepts of a same color belong to the same equivalence class according to the chosen congruence. Right: quotient lattice.

The quotient lattice $\mathbb{C}/\theta$ formalizes the partition of the harmonic system, preserving its structure. A sub-lattice of $\mathbb{C}(\mathbb{K})$ isomorphic to $\mathbb{C}/\theta$ is interpreted as a harmonic sub-system compatible with the harmonic structure generated by the partition defined from a set of harmonic forms.

Example 3. An interesting congruence, from a musical point of view, can be defined by gathering the most common harmonic tonal forms in the same equivalence class (perfect chords, seventh chords and ninth chords). The generating elements for this class are $\{0, 2, 4\}$, $\{0, 1, 3, 5\}$, and $\{0, 1, 2, 3, 5\}$ (again only the extent is mentioned here). Another class is generated from $\{0, 3\}$ (i.e. the fourths, which are also interesting from a musical point of view). The other classes are derived to preserve the ordering relations. This congruence θ^* is illustrated in Fig. 3, still for the 7-tet, along with the corresponding quotient lattice. The first generated class is displayed in pink and the second one in green. In this case, the quotient lattice is simply a chain, representing a linear complete ordering among equivalence classes.[7] The remaining concepts (excluding the top and the bottom of the lattice) form a third equivalence class. Let us consider the following example, to illustrate the consistency of the generated classes: $\{0,3\}$ and $(\{0,1,3\},\{0,2,3\})$ are congruent (both are in the green class); similarly $\{0,1,3,5\}$ and $\{0,1,2,3\}$ are congruent (both in the pink class). The conjunctions $\{0,3\} \wedge \{0,1,3,5\} = \{0,3\}$ and $(\{0,1,3\},\{0,2,3\}) \wedge \{0,1,2,3\} = (\{0,1,3\},\{0,2,3\})$ are congruent, and the disjunctions $\{0,3\} \vee \{0,1,3,5\} = \{0,1,3,5\}$ and $(\{0,1,3\},\{0,2,3\}) \vee \{0,1,2,3\} = \{0,1,2,3\}$ are also congruent.

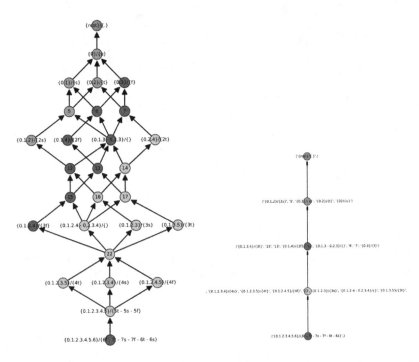

Fig. 3. Left: congruence θ^* on the 7-tet concept lattice. Right: quotient lattice $\mathbb{C}/\theta^*$.

We now propose to exploit the two notions of congruence and of morphological operators to define musical descriptors.

Definition 10 (Harmonico-morphological descriptors). *Let $\mathcal{M}$ be a musical piece, $\mathbb{T}_{\mathcal{M}}$ the harmonic system associated with it, and $\mathbb{C}(\mathcal{M})$ the corresponding concept lattice. The core idea of the proposed descriptors is to use dilations and erosions of the set of formal concepts to provide upper and lower bounds of the description on the one hand, and congruences to provide a structural*

---

[7] An interesting question, which still remains open, concerns the possible ways of generating chains which are musically relevant by carefully selecting the underlying equivalence classes.

summary of the harmonic forms on the other hand. The set of formal concepts corresponding to the harmonic forms in $\mathcal{M}$ is denoted by $H_{\mathbb{C}}^{\mathcal{M}}$. The dilations δ and erosions ε can typically be defined from a metric associated with a valuation on $\mathbb{C}(\mathcal{M})$. Three congruences are then defined:

- *θ grouping all formal concepts in $H_{\mathbb{C}}^{\mathcal{M}}$ into one same class;*
- *θ_δ grouping all formal concepts in $\delta(H_{\mathbb{C}}^{\mathcal{M}})$ into one same class;*
- *θ_ε grouping all formal concepts in $\varepsilon(H_{\mathbb{C}}^{\mathcal{M}})$ into one same class.*

The proposed harmonic descriptors are the quotient lattices $\mathbb{C}(\mathcal{M})/\theta$, $\mathbb{C}(\mathcal{M})/\theta_\delta$, and $\mathbb{C}(\mathcal{M})/\theta_\varepsilon$.

We argue that these descriptors are good representative of $\mathcal{M}$, since they preserve the intervallic structures, and provide compact summaries, which would allow for comparison between musical pieces.

4.2   Algorithm

The procedure developed to generate the proposed descriptors is given in Algorithm 1 (the implementation was done in SageMath).

Algorithm 1. Generating a morphological interval of harmonic descriptors based on congruence relations and quotient lattices

---

Require: $\mathbb{C}(\mathcal{M})$: concept lattice built from the harmonic system $\mathbb{T}_{\mathcal{M}}$
Require: $H^{\mathcal{M}}$: set of harmonic forms present in $\mathcal{M}$
Require: w: valuation on $\mathbb{C}(\mathcal{M})$
Require: $\mathbb{M}_w$: metric associated with w on $\mathbb{C}(\mathcal{M})$
Require: (ε, δ): adjunction $\mathbb{C}(\mathcal{M})$, where the dilation δ and the erosion ε are built from the metric $\mathbb{M}_w$
Require: n size of the dilation and erosion
Ensure: $\mathbb{C}(\mathcal{M})/\theta$, $\mathbb{C}(\mathcal{M})/\theta_\delta$ et $\mathbb{C}(\mathcal{M})/\theta_\varepsilon$: harmonico-morphological descriptors of $\mathcal{M}$
1: **function** HARMONICO-MORPHOLOGICAL DESCRIPTORS($\mathbb{C}(\mathcal{M})$, $H^{\mathcal{M}}$, (δ, ε), n)
2:     Compute the set of formal concepts $H_{\mathbb{C}}^{\mathcal{M}}$ associated with the harmonic forms present in $H^{\mathcal{M}}$
3:     Compute the congruence relation θ on $\mathbb{C}(\mathcal{M})$ such that all concepts in $H_{\mathbb{C}}^{\mathcal{M}}$ belong to the same equivalence class $[\cdot]_\theta$
4:     Compute the dilation $\delta(H_{\mathbb{C}}^{\mathcal{M}})$ and the erosion $\varepsilon(H_{\mathbb{C}}^{\mathcal{M}})$ of the set of formal concepts $H_{\mathbb{C}}^{\mathcal{M}}$ using a structuring element defined as a ball of radius n of the considered metric
5:     Compute the congruence relation θ_δ such that all concepts in $\delta(H_{\mathbb{C}}^{\mathcal{M}})$ belong to the same equivalence class $[\cdot]_{\theta_\delta}$
6:     Compute the congruence relation θ_ε such that all concepts in $\varepsilon(H_{\mathbb{C}}^{\mathcal{M}})$ belong to the same equivalence class $[\cdot]_{\theta_\varepsilon}$
7:     Compute the quotient lattices $\mathbb{C}(\mathcal{M})/\theta$, $\mathbb{C}(\mathcal{M})/\theta_\delta$ and $\mathbb{C}(\mathcal{M})/\theta_\varepsilon$ according to the congruence relations θ, θ_δ and θ_ε, respectively
8:     **return** $(\mathbb{C}(\mathcal{M})/\theta, \mathbb{C}(\mathcal{M})/\theta_\delta, \mathbb{C}(\mathcal{M})/\theta_\varepsilon)$

4.3   Example on Ligeti's String Quartet No. 2

As an illustrative example, we apply the proposed method for computing the musical descriptors on $\mathcal{M}$ corresponding to the first movement of Ligeti's String Quartet No. 2. For example, the first set, $\{0, 1\}$, corresponds to the two notes chord $\{C, D\}$, whereas the last one, $\{0, 1, 2, 3\}$, corresponds to the tetrachord $\{C, D, E, F\}$.

We use the previous 7-tet lattice for the analysis by selecting a limited number of harmonic forms which are used by the composer. These forms are given in Fig. 4 by means of a circular representation corresponding to the underlying cyclic group of order 7.

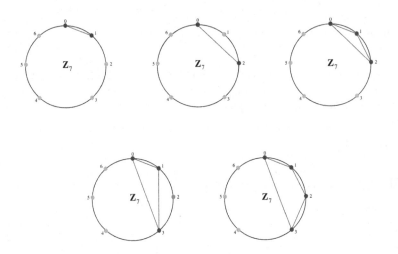

Fig. 4. Some harmonic forms in Ligeti's quartet fragment.

The chosen valuation is the cardinality of the filter w_F (see Sect. 3). Note that other valuations could be used as well. The associated distance was used to defined elementary dilations and erosions (with $n = 2$) on join-irreducible (respectively meet-irreducible) elements. The dilation or erosion of any concept is then derived from its decomposition into irreducible elements and using the commutativity with the supremum, respectively the infimum.

The steps of Algorithm 1 for this fragment are illustrated in Figs. 5, 6 and 7. As we could guess from the congruence relations, the final quotient lattices show isomorphic relations between $\mathbb{C}(\mathcal{M})/\theta$ and $\mathbb{C}(\mathcal{M})/\theta_\delta$. The larger number of different congruence classes in the erosion $\varepsilon(H_{\mathbb{C}}^{\mathcal{M}})$ is reflected in the form of the corresponding quotient lattice $\mathbb{C}(\mathcal{M})/\theta_\varepsilon$, which contains more elements.

This example is particularly interesting because the musical excerpt does not only contain the usual perfect major and minor, seventh and ninth chords. However, the use of the 7-tet, used here for the simplicity of the illustration, is too limited. It would be even more interesting to use the 12-tet, which would better account for chromatic parts. This is surely more relevant for musical pieces where the chromaticism is more relevant than the diatonic component. However, the proposed approach paves the way for such deeper investigations.

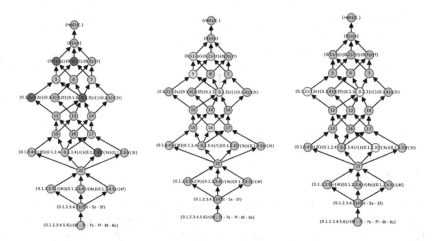

Fig. 5. Formal concepts associated with the harmonic forms found in $H^{\mathcal{M}}$. From left to right: $H_{\mathbb{C}}^{\mathcal{M}}$ (concepts displayed in red), dilation $\delta(H_{\mathbb{C}}^{\mathcal{M}})$ (green concepts), and erosion $\varepsilon(H_{\mathbb{C}}^{\mathcal{M}})$ (yellow concepts). (Color figure online)

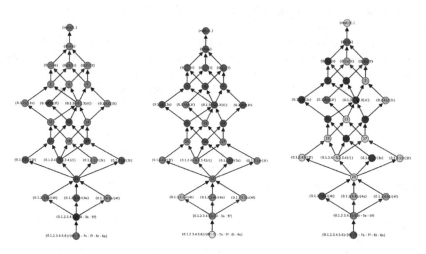

Fig. 6. Congruence relations θ, θ_δ, and θ_ε on $\mathbb{C}(\mathcal{M})$ (7-tet) generated by: $H_{\mathbb{C}}^{\mathcal{M}}$, $\delta(H_{\mathbb{C}}^{\mathcal{M}})$, and $\varepsilon(H_{\mathbb{C}}^{\mathcal{M}})$, respectively (from left to right).

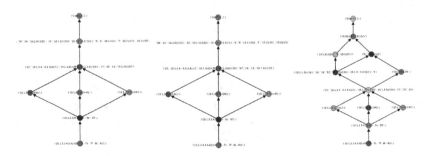

Fig. 7. Quotient lattices. From left to right: $\mathbb{C}(\mathcal{M})/\theta$, $\mathbb{C}(\mathcal{M})/\theta_\delta$, and $\mathbb{C}(\mathcal{M})/\theta_\varepsilon$.

5   Conclusion

This paper suggests for the first time a possible strategy to approach music information research by combining tools and ideas belonging to two autonomous fields, i.e. Mathematical Morphology and Formal Concept Analysis. Although some of the concepts described in the paper had already found potential applications in music analysis, it is the first attempt at conceiving structural descriptors based on the joint exploitation of these concepts within a computational musicological perspective. Introducing congruence relations in lattice-based representations provides a new way of extracting and summarizing the information contained in a musical piece by preserving the core intervallic structure. The proposed descriptors, particularly suited for atonal and contemporary music which explores the whole space of harmonic forms, are aimed to be used to characterize styles of music, or comparing different pieces of music, using matching between quotient lattices for instance.

Further investigations are needed in order to find meaningful distances (or pseudo-distances) between formal concepts in order to express musically relevant morphological operations. At a more abstract level, the question of comparing quotient lattices still remains open in the context of music information research. This goes with the definition of similarities between the descriptors, as for example by establishing whether quotient lattices from two different pieces, or two excerpts of a musical piece, are (or are not) isomorphic. In order to make this comparison computationally reasonable, the compact representation provided by the quotient lattice would be directly exploited. This would clearly provide an assessment of the structural and harmonic similarity between them.

References

1. Atif, J., Bloch, I., Distel, F., Hudelot, C.: Mathematical morphology operators over concept lattices. In: Cellier, P., Distel, F., Ganter, B. (eds.) ICFCA 2013. LNCS (LNAI), vol. 7880, pp. 28–43. Springer, Heidelberg (2013). https://doi.org/10.1007/978-3-642-38317-5_2
2. Atif, J., Bloch, I., Hudelot, C.: Some relationships between fuzzy sets, mathematical morphology, rough sets, F-transforms, and formal concept analysis. Int. J. Uncertainty Fuzziness Knowl.-Based Syst. **24**(S2), 1–32 (2016)

3. Barbut, M., Frey, L., Degenne, A.: Techniques ordinales en analyse des données. Hachette, Paris (1972)
4. Birkhoff, G.: Lattice Theory, vol. 25, 3rd edn. American Mathematical Society, Providence (1979)
5. Bloch, I., Heijmans, H., Ronse, C.: Mathematical morphology. In: Aiello, M., Pratt-Hartman, I., van Benthem, J. (eds.) Handbook of Spatial Logics, pp. 857–947. Springer, Dordrecht (2007). https://doi.org/10.1007/978-1-4020-5587-4_14
6. Ganter, B., Wille, R.: Formal Concept Analysis. Springer, Heidelberg (1999). https://doi.org/10.1007/978-3-642-59830-2
7. Heijmans, H.J.A.M.: Morphological Image Operators. Academic Press, Boston (1994)
8. Heijmans, H.J.A.M., Ronse, C.: The algebraic basis of mathematical morphology - part I: dilations and erosions. Comput. Vis. Graph. Image Proc. **50**, 245–295 (1990)
9. Leclerc, B.: Lattice valuations, medians and majorities. Discret. Math. **111**(1), 345–356 (1993)
10. Monjardet, B.: Metrics on partially ordered sets–a survey. Discret. Math. **35**(1), 173–184 (1981)
11. Najman, L., Talbot, H.: Mathematical Morphology: From Theory to Applications. ISTE-Wiley, Hoboken (2010)
12. Neumaier, W., Wille, R.: Extensionale Standardsprache der Musiktheorie: eine Schnittstelle zwischen Musik und Informatik. In: Hesse, H.P. (ed.) Mikrotöne III, pp. 149–167. Helbing, Innsbruck (1990)
13. Noll, T., Brand, M.: Morphology of chords. In: Perspectives in Mathematical and Computational Music Theory, vol. 1, p. 366 (2004)
14. Ronse, C.: Adjunctions on the lattices of partitions and of partial partitions. Appl. Algebra Eng. Commun. Comput. **21**(5), 343–396 (2010)
15. Ronse, C., Heijmans, H.J.A.M.: The algebraic basis of mathematical morphology - part II: openings and closings. Comput. Vis. Graph. Image Proc. **54**, 74–97 (1991)
16. Schlemmer, T., Andreatta, M.: Using formal concept analysisto represent chroma systems. In: Yust, J., Wild, J., Burgoyne, J.A. (eds.) MCM 2013. LNCS (LNAI), vol. 7937, pp. 189–200. Springer, Heidelberg (2013). https://doi.org/10.1007/978-3-642-39357-0_15
17. Schlemmer, T., Schmidt, S.E.: A formal concept analysis of harmonic forms and interval structures. Ann. Math. Artif. Intell. **59**(2), 241–256 (2010)
18. Serra, X., Magas, M., Benetos, E., Chudy, M., Dixon, S., Flexer, A., Gómez, E., Gouyon, F., Herrera, P., Jorda, S., et al.: Roadmap for music information research (2013)
19. Serra, J. (ed.): Image Analysis and Mathematical Morphology, Part II: Theoretical Advances. Academic Press, London (1988)
20. Wille, R.: Restructuring lattice theory: an approach based on hierarchies of concepts. In: Rival, I. (ed.) Ordered Sets. ASIC, pp. 445–470. Springer, Dordrecht (1982). https://doi.org/10.1007/978-94-009-7798-3_15
21. Wille, R.: Sur la fusion des contextes individuels. Mathématiques et Sciences Humaines **85**, 57–71 (1984)
22. Wille, R.: Musiktheorie und Mathematik. In: Götze, H. (ed.) Musik und Mathematik, pp. 4–31. Springer, Heidelberg (1985). https://doi.org/10.1007/978-3-642-95474-0_2
23. Wille, R.: Restructuring lattice theory: an approach based on hierarchies of concepts. In: Ferré, S., Rudolph, S. (eds.) ICFCA 2009. LNCS (LNAI), vol. 5548, pp. 314–339. Springer, Heidelberg (2009). https://doi.org/10.1007/978-3-642-01815-2_23

Towards Collaborative Conceptual Exploration

Tom Hanika[1,2(✉)] and Jens Zumbrägel[3]

[1] Knowledge and Data Engineering Group, University of Kassel, Kassel, Germany
tom.hanika@cs.uni-kassel.de
[2] Interdisciplinary Research Center for Information System Design,
University of Kassel, Kassel, Germany
[3] Faculty of Computer Science and Mathematics,
University of Passau, Passau, Germany
jens.zumbraegel@uni-passau.de

Abstract. In domains with high knowledge distribution a natural objective is to create principle foundations for collaborative interactive learning environments. We present a first mathematical characterization of a collaborative learning group, a *consortium*, based on closure systems of attribute sets and the well-known attribute exploration algorithm from formal concept analysis. To this end, we introduce (weak) local experts for subdomains of a given knowledge domain. These entities are able to refute and potentially accept a given (implicational) query for some closure system that is a restriction of the whole domain. On this we build up a *consortial expert* and show first insights about the ability of such an expert to answer queries. Furthermore, we depict techniques on how to cope with falsely accepted implications and on combining counterexamples. Using notions from combinatorial design theory we further expand those insights as far as providing first results on the decidability problem if a given consortium is able to explore some target domain. Applications in conceptual knowledge acquisition as well as in collaborative interactive ontology learning are at hand.

Keywords: Formal concept analysis · Implications
Attribute exploration · Collaborative knowledge acquisition
Collaborative interactive learning

1 Introduction

Collaborative knowledge bases, like DBpedia[1] and Wikidata[2] [16], raise the need for (interactive) collaborative tools in order to add, enhance or extract conceptual knowledge to and from those. As well, a society with highly specialized experts needs some method to make use of the collaborative knowledge of those.

The authors are given in alphabetical order. No priority in authorship is implied.
[1] http://wiki.dbpedia.org.
[2] http://www.wikidata.org.

© Springer International Publishing AG, part of Springer Nature 2018
P. Chapman et al. (Eds.): ICCS 2018, LNAI 10872, pp. 120–134, 2018.
https://doi.org/10.1007/978-3-319-91379-7_10

One particular task in knowledge acquisition is to obtain concepts in a given domain which is composed of two disjoint sets, called *objects* and *attributes*, along with some relation between them. A well-known approach for this is the (classical) attribute exploration algorithm from formal concept analysis (FCA) [3,5]. This algorithm is able to explore any domain of the kind mentioned above by consulting some domain expert. The result is a formal concept lattice, i.e., an order-theoretic lattice which contains all formal concepts discovered in the domain. It is crucial that the algorithm has access to a *domain expert* for the whole domain, to whom it uses a minimal number of queries (which may still be exponential in the size of input, i.e., the size of the relation between objects and attributes).

However, the availability of a domain expert is often not given in practice. Moreover, even if it exists, such an expert might not be able or willing to answer the possibly exponential number of queries. The purpose of the present work is to provide a solution in this case, at least for some of such tasks, given a certain collaborative scenario. More precisely, suppose that we have a covering $M = \bigcup_{i \in I} N_i$ of the attribute set M together with a set of *local experts* p_i on N_i, then we propose a *consortial expert* for the domain. As is easy to see, such an expert is in general less capable of handling queries than a domain expert. Nonetheless, depending on the form of $\mathcal{M} = \{N_i \mid i \in I\}$ our approach may still be able to answer a significant amount of non-trivial queries.

In this work we provide a first complete characterization of (weak) local experts in order to define what a *consortium* is, what can be explored and what next steps should be focused on. As to our knowledge, this has not been considered before in the realm of conceptual knowledge.

Here is an outline of the remainder of this paper. After giving an account of related work in Sect. 2, we recall basic notions from formal concept analysis and the attribute exploration algorithm in Sect. 3. We define the setting of a consortium in Sect. 4, using a small simplification in notation to mere closure systems on M. Subsequently we discuss our approach in Sect. 5, give examples in Sect. 5.3, following by possible extensions in Sect. 5.4 and a conclusion in Sect. 6.

2   Related Work

There are several related fields that address the problem of (interactive) collaborative learning in their respective scientific languages. Based on modal logic there are various new approaches for similar problems as considered here, using epistemic and intuitionistic types. For example, Jäger and Marti [7] present a multi-agent system for intuitionistic distributed knowledge (with truth). Another example is resolving the distributed knowledge of a group as done by Ågotnes and Wáng [1]. In this work the process of *distributed knowledge*, i.e., knowledge distributed throughout a group, is resolved to common knowledge, i.e., knowledge that is known to all members of the group, a fact which is also known to the members of the group.

Investigations considering a more virtual approach for collaborative knowledge acquisition are, for example, presented by Stange et al. [13], in which a

collaborative graphical editor used by experts negotiates the outcome. Our approach is yet based on (basic) formal attribute exploration [5]. Of course, there
are various advanced versions like adding background knowledge [3], relational
exploration [12] or conceptual exploration [14]. There are also extensions of the
basic exploration to treat incomplete knowledge [2,6,10].

In FCA one of the first considerations on cooperatively building knowledge
bases is work of Martin and Eklund [9]. Previous work on collaborative interactive concept lattice modification in order to extract knowledge can be found
in [15]. These concept lattice modifications are based on removing or adding
attributes/objects/concepts using expert knowledge, and those operations may
be used in a later version of collaborative conceptual exploration. The most
recent work specifically targeting collaborative exploration is [11], raising the
task of making exploration collaborative.

3   Attribute Exploration and FCA Basics

In this paper we utilize notions from formal concept analysis (FCA) as specified
in [5]. In short, our basic data structure is a *formal context* $\mathbb{K} := (G, M, I)$
with G some object set, M some attribute set, and $I \subseteq G \times M$ an incidence
relation between them. By $\cdot'$ we denote two mappings $\cdot': \mathcal{P}(G) \to \mathcal{P}(M)$ and
$\cdot': \mathcal{P}(M) \to \mathcal{P}(G)$, given by $A \mapsto A' = \{m \in M \mid \forall g \in A : (g, m) \in I\}$ for $A \subseteq G$
and $B \mapsto B' = \{g \in G \mid \forall m \in B : (g, m) \in I\}$ for $B \subseteq M$.

The set $\mathfrak{B}(\mathbb{K})$ is the set of all *formal concepts*, i.e., the set of all pairs (A, B)
with $A \subseteq G$, $B \subseteq M$ such that $A' = B$ and $B' = A$. In a formal concept (A, B)
the set A is called (concept-)extent and the set B is called (concept-)intent. The
set of all formal concepts can be ordered by $(A, B) \leq (C, D) :\Leftrightarrow A \subseteq C$. The
ordered set $\mathfrak{B}(\mathbb{K})$, often denoted by $\underline{\mathfrak{B}}(\mathbb{K})$, is called the *concept lattice* of $\mathbb{K}$.
Furthermore, the composition $\cdot''$ constitutes closure operators on G and on M,
respectively, i.e., mappings $\cdot'': \mathcal{P}(G) \to \mathcal{P}(G)$ and $\cdot'': \mathcal{P}(M) \to \mathcal{P}(M)$ which are
extensive, monotone and idempotent. Therefore, every formal context gives rise,
through the associated closure operator, to two closure systems, one on G and
one on M, called the closure system of intents and extents, respectively. Each
of those closure systems can be considered as an ordered set using the inclusion
operator $\subseteq$, which in turn leads to a complete lattice. Using the basic theorem of
FCA [5] one may construct for any closure system $\mathcal{X}$ on M a formal context $\mathbb{K}$
such that the closure system $\mathcal{X}$ is the set of concept-intents from $\mathbb{K}$.

In the following exposition we will concentrate on the attribute set M of a
formal context. We do this for brevity and clarity reasons, only. Namely, we avoid
carrying all the necessary notation through the defining parts of a collaborating
consortium. However, we do keep in mind that M is still a part of a formal
context (G, M, I), and we rest on this classical representation, in particular,
when quoting well-known algorithms from FCA.

3.1   Setting

Let M be some finite (attribute) set. We fix a closure system $\mathcal{X} \subseteq \mathcal{P}(M)$, called the *(target) domain* or *target closure system*, which is the domain knowledge to be acquired. The set of all closure systems on a set M constitutes a closure system itself. In turn, this means we can also find a concept lattice for this set. We depict this lattice in general in Fig. 1 (right). The size of this set is enormous and only known up to $|M| = 7$. In the next subsection we recall the classical algorithm to compute the target domain for a given set of attributes M using a domain expert on M. This algorithm employs rules between sets of attributes which we now recall. An *implication* is a pair $(A, B) \in \mathcal{P}(M) \times \mathcal{P}(M)$, which can also be denoted by $A \rightarrow B$. We write $\text{Imp}(M)$ for the set $\mathcal{P}(M) \times \mathcal{P}(M)$ of all implications on M. The implication $(A, B) \in \text{Imp}(M)$ is *valid* in $\mathcal{X}$ if $\forall X \in \mathcal{X} : A \subseteq X \Rightarrow B \subseteq X$.

3.2   Attribute Exploration

Attribute exploration is an instance of an elegant strategy to explore the knowledge of an (unknown) domain (G, M, I) using queries to a domain expert for M. These queries consist of validity questions concerning implications in M. The expert in this setting can either accept an implication, i.e., confirming that this implication is valid in the domain, or has to provide a counterexample. The following description of this algorithm is gathered from [4], a compendium on conceptual exploration methods.

Using a signature, which specifies the logical language to be used during exploration, there is a set of possible implications $\mathcal{F}$, each either valid or not in the domain. The algorithm itself uses an exploration knowledge base $(\mathcal{L}, \mathcal{E})$, with $\mathcal{L}$ being the set of the already accepted implications and $\mathcal{E}$ the set of already collected counterexamples. These can be considered in our setting as named subsets of M, where the name is the object name for this set. The algorithm now makes use of a query engine which draws an implication f from $\mathcal{F}$ that cannot be deduced from $\mathcal{L}$ and that cannot be refuted by already provided counterexamples in $\mathcal{E}$. This implication is presented to the domain expert, who either can accept this implication, which adds f to $\mathcal{L}$, or refute f by a counterexample $E \subseteq M$, which adds E to $\mathcal{E}$.

The crucial part here is that the domain expert has to be an expert for the whole domain, i.e., an expert for the whole attribute set M and any object possible. Otherwise, the expert would not be able to provide complete counterexamples, i.e., the provided counterexamples are possibly missing attributes from M, or even "understand" the query. To deal with this impractical limitation algorithms for attribute exploration with partial (counter-)examples were introduced. We refer the reader to [4, Algorithm 21]. This algorithm is able to accept partial counterexamples from a domain expert.

The return value of the attribute exploration algorithm is the canonical base of all valid implications from the domain. There is no smaller set of implications than the canonical base for some closure operator on an (attribute) set M, which is sound and complete.

In the subsequent section we provide a characterization of a consortial expert which could be utilized as such a domain expert providing incomplete counterexamples. In addition, we show a strategy for how to deal with counterexamples de-validating already accepted implications, which will be a possible outcome when consulting a consortium.

4   Consortium

In the following we continue to utilize mere closure systems on M for some domain (G, M, I) and also call such a closure system itself the (target) domain $\mathcal{X}$, to be explored. This ambiguity is for brevity, only. Furthermore, we consider M always to be finite.

Definition 4.1 (Expert). An *expert* for $\mathcal{X}$ is a mapping $p \colon \mathrm{Imp}(M) \to \mathcal{X} \cup \{\top\}$ such that for every $f = (A, B) \in \mathrm{Imp}(M)$ the following is true:

(1) $p(f) = \top \Rightarrow f$ is valid in $\mathcal{X}$,
(2) $p(f) = X \in \mathcal{X} \Rightarrow A \subseteq X \wedge B \not\subseteq X$.

We refer to the set M also as the *expert domain*.

From this definition we note, for an implication $f \in \mathrm{Imp}(M)$, that $p(f) \neq \top$ implies that f is not valid in $\mathcal{X}$, since $p(f) = X \in \mathcal{X} \Rightarrow A \subseteq X \wedge B \not\subseteq X$. In analogy to this expert we now introduce an expert on a subset of M.

Definition 4.2 (Local expert). Let $N \subseteq M$. A *local expert* for $\mathcal{X}$ on N is a mapping $p_N \colon \mathrm{Imp}(N) \to \mathcal{X}_N \cup \{\top\}$ with $\mathcal{X}_N := \{X \cap N \mid X \in \mathcal{X}\}$ such that for every $f = (A, B) \in \mathrm{Imp}(N)$ there holds:

(1) $p_N(f) = \top \Rightarrow f$ is valid in $\mathcal{X}$,
(2) $p_N(f) = X \in \mathcal{X}_N \Rightarrow A \subseteq X \wedge B \not\subseteq X$.

Observe that the set $\mathcal{X}_N$ is also a closure system. Despite that, the elements of $\mathcal{X}_N$ are not necessarily elements of $\mathcal{X}$. But, since $N \subseteq M$ there is for every $X \in \mathcal{X}_N$ some $\hat{X} \in \mathcal{X}$ such that $\hat{X} \cap N = X$.

Remark 4.3. Every expert for $\mathcal{X}$ provides in the obvious way a local expert for $\mathcal{X}$ on N, for each $N \subseteq M$. Furthermore, every local expert for $\mathcal{X}$ on N is a local expert for $\mathcal{X}$ on O for each $O \subseteq N$.

Lemma 4.4 (Refutation by local expert). *Let $\mathcal{X}$ be some domain with attribute set M and let p_N be a local expert for $\mathcal{X}$ on $N \subseteq M$. Then for every $f \in \mathrm{Imp}(N)$ there holds $p_N(f) \neq \top \Rightarrow f$ is not valid in $\mathcal{X}$.*

Proof. If $p_N(f) \neq \top$, then $\exists X \in \mathcal{X}_N : p_N(f) = X \wedge A \subseteq X \wedge B \not\subseteq X$. By definition $\exists \hat{X} \in \mathcal{X} : \hat{X} \cap N = X$. Therefore, $A \subseteq X = \hat{X} \cap N \subseteq \hat{X}$ and $B \not\subseteq X = \hat{X} \cap N$, thus $B \not\subseteq \hat{X}$ as $B \subseteq N$. □

Example 4.5. Suppose we have a three-element attribute set $M = \{ro, fl, ed\}$, for the attributes "round", "flexible" and "edible". Regarding the objects "ball", "sphere" and "donut" (food) we consider the following formal context.

	round	flexible	edible
ball	✕	✕	
sphere	✕		
donut		✕	✕

From this we obtain as our target domain

$$\mathcal{X} = \big\{ M, \{ro, fl\}, \{fl, ed\}, \{fl\}, \{ro\}, \varnothing \big\},$$

with the canonical base $\mathcal{B} = \{ed \rightarrow fl\}$. Using the shortcuts $ed^C = \{ro, fl\}$ and $ro^C = \{fl, ed\}$, the concept lattice may be depicted as:

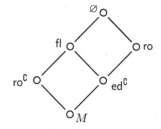

Now suppose that $I = \{a, b, c\}$, and for each $i \in I$ we have a local expert p_i for $\mathcal{X}$ on N_i, where $N_a = \{ro, fl\}$, $N_b = \{fl, ed\}$ and $N_c = \{ro, ed\}$. We name the local experts "Alice", "Bob" and "Carol".

Alice may be consulted for the implications $ro \rightarrow fl$ and $fl \rightarrow ro$, both of which she refutes. For example, to the query $ro \rightarrow fl$ she responds (possibly having the sphere in mind) with an attribute set X containing ro but not fl, i.e., $X = \{ro\}$, where $\{ro\} = \hat{X} \cap \{ro, fl\}$ and $\hat{X} \in \mathcal{X}$. Similarly, she refutes the query $fl \rightarrow ro$ (having the donut in mind). Moreover, local expert Bob can be consulted with the implications $fl \rightarrow ed$, which he refutes (ball), and $ed \rightarrow fl$, which he correctly accepts. Finally, Carol refutes both possible queries $ed \rightarrow ro$ (donut) and $ro \rightarrow ed$, in which case her counterexample could stem from different objects (ball or sphere).

For some applications a local expert may be too strong in terms of having either to accept an implication (vicariously for $\mathcal{X}$) or refute an implication. This would require that the local expert is aware of all possible counterexamples, which is impractical.

Definition 4.6 (Local pre-expert). A *local pre-expert* for $\mathcal{X}$ on $N \subseteq M$ is a mapping $p_N^* \colon \mathrm{Imp}(N) \rightarrow \mathcal{X}_N \cup \{\top\}$ such that $\forall f = (A, B) \in \mathrm{Imp}(N) \colon p_N^*(f) = X \in \mathcal{X}_N \Rightarrow A \subseteq X \land B \not\subseteq X$.

It is obvious that a local expert is also a local pre-expert. Using this "weaker" mapping we introduce the consortial (pre-)expert, after stating what a consortial domain is and some technical result about the intersection of closed sets.

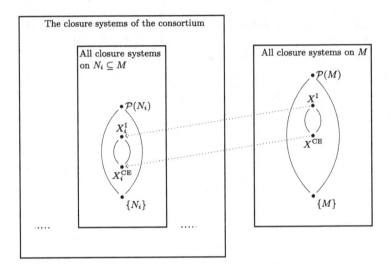

Fig. 1. Closure system of all closure systems on M (right) and on $N_i \subseteq M$ (left). The closure systems for the set of accepted implications are denoted by X^I (in M) and by X_i^I (in N_i), and likewise for the set of counterexamples by X^{CE} and by X_i^{CE}.

Definition 4.7 (Consortial domain). Let M be some attribute set and $\mathcal{X} \subseteq \mathcal{P}(M)$ be the target domain. Then a family $\mathcal{M} = \{N_i \mid i \in I\} \subseteq \mathcal{P}(M)$ for some index set I is called *consortial domain on* M if $\bigcup_{i \in I} N_i = M$.

We call $\mathcal{M} \subseteq \mathcal{P}(M)$ a *proper consortial domain* if $M \notin \mathcal{M}$.

Lemma 4.8 (Consortial domain closed under intersection). *Let $\mathcal{M}$ be some consortial domain on M. If $\mathcal{M}$ is closed under intersection, then so is the set $\bigcup_{M \in \mathcal{M}} \mathcal{X}_M$.*

In the following proof as well as in the rest of this work we may often use the abbreviation $\mathcal{X}_i := \mathcal{X}_{N_i}$ for some $N_i \in \mathcal{M}$ with $\mathcal{M}$ a consortial domain and using the notation introduced in Definition 4.2.

Proof. Whenever $X \cap N_i \in \mathcal{X}_i$ and $Y \cap N_j \in \mathcal{X}_j$, where $X, Y \in \mathcal{X}$, we get

$$(X \cap N_i) \cap (Y \cap N_j) = (X \cap Y) \cap (N_i \cap N_j) \in \mathcal{X}_{N_i \cap N_j},$$

where $X \cap Y \in \mathcal{X}$ and $N_i \cap N_j \in \mathcal{M}$. □

Corollary 4.9. *If $\mathcal{M}^* := \mathcal{M} \cup \{M\}$ is a closure system, then so is $\bigcup_{M \in \mathcal{M}^*} \mathcal{X}_M$.*

By definition a proper consortial domain cannot be a closure system and even a consortial domain will almost never have this property, either. However, for any consortial domain $\mathcal{M}$ we can easily construct an intersection closed set using the downset operator $\downarrow \mathcal{M}$. Therefore, whenever we have a consortial domain we may consider $\downarrow \mathcal{M}$, when necessary. Hence, we always can construct a closure system $\mathcal{M}^*$ for any consortial domain $\mathcal{M}$.

In the following we may use M^* to speak about $\downarrow \mathcal{M} \cup \{M\}$.

Remark 4.10 (Closure operator $\mathcal{M}^*$). Since for a given consortial domain $\mathcal{M}$ on M the set $\mathcal{M}^*$ is a closure system, we obtain a closure operator $\phi : \mathcal{P}(M) \to \mathcal{P}(M)$. We may address ϕ simply by $\mathcal{M}^*()$ and the image of $N \subseteq M$ by $\mathcal{M}^*(N)$.

Using the just discovered closure operator we may define which queries can be answered in a consortial domain.

Definition 4.11 (Well-formed query). Let $\mathcal{M}$ be some proper consortial domain on M and let $f = (A, \{b\}) \in \mathrm{Imp}(M)$. Then f is called *well-formed* for $\mathcal{M}$ if $\mathcal{M}^*(A \cup \{b\}) \neq M$, i.e., if there exists $N_i \in \mathcal{M}$ such that $A \cup \{b\} \subseteq N_i$.

Well-formed queries are in fact the only queries for which in a proper consortial domain the decision problem if an implication is valid can be decided. It is easy to see that for any given $f = (A, \{b\}) \in \mathrm{Imp}(M)$, if $\mathcal{M}^*(A \cup \{b\}) = M$, then there is no expert domain left, therefore either the conclusion attribute or one of the premises is missing in all $N \in \mathcal{M}$, which leads to undecidability.

Putting all those ideas together we are finally able to define our main goal.

Definition 4.12 (Consortium for $\mathcal{X}$). For an attribute set M and a target domain $\mathcal{X}$ on M let $\mathcal{M} = \{N_i \mid i \in I\}$ be a consortial domain on M. A *consortium* for $\mathcal{X}$ is a family $\mathcal{C} := \{p_i\}_{i \in I}$ of local pre-experts p_i for $\mathcal{X}$ on N_i.

All comments made before about $\mathcal{M}$ being intersection-closed are compatible with the definition of a consortium. Using Remark 4.3 we can always obtain a local pre-expert for any $\hat{M} \in \downarrow \mathcal{M}$. A consortium is able to decide the validity of any well-formed query, by definition. Therefore, a consortium gives rise to a *consortial expert*. As long as all queries are well-formed, a consortium can be used in-place of a domain expert.

Example 4.13. We continue with Example 4.5. On the consortial domain $\mathcal{M} := \{N_a, N_b, N_c\}$ the three local experts form a consortium $\mathcal{C} := \{p_a, p_b, p_c\}$ for $\mathcal{X}$. Note that the consortium cannot decide, e.g., on the implication $\{\mathsf{fl}, \mathsf{ed}\} \to \mathsf{ro}$, since this query is not well-formed for $\mathcal{M}$. However, if experts are able to combine their counterexamples they may refute the query (cf. Sect. 5.4).

Definition 4.14 (Strong consortial expert). Let $\mathcal{C} = \{p_i\}_{i \in I}$ be a consortium for $\mathcal{X}$ on M. A *strong consortial expert* is a mapping $p_\mathcal{C} : \bigcup_{i \in I} \mathrm{Imp}(N_i) \to \bigcup_{i \in I} \mathcal{X}_i \cup \{\top\}$ such that for every $f = (A, B) \in \bigcup_{i \in I} \mathrm{Imp}(N_i)$ there holds:

(1) $\exists p_i \in \mathcal{C}, p_i(f) \neq \top \Rightarrow p_\mathcal{C}(f) \neq \top$,
(2) $p_\mathcal{C}(f) = X \in \bigcup_{i \in I} \mathcal{X}_i \Rightarrow A \subseteq X \wedge B \not\subseteq X$.

The strong consortial expert has to respect a possible counterexample entailed in the consortium in order to be consistent with Definition 4.6, since every counterexample by a local (pre-)expert is a restriction of an element of the target closure system. In the case of having local experts in the consortium this behavior may be in conflict with Definition 4.2, since we demand that accepting an implication by a local expert implies that the implication is true in the target domain. For example, if a local expert accepts an implication and another local (pre-)expert refutes it, this conflict is not resolvable. Therefore, whenever a consortium does contain local experts it is mandatory that they meet a consistency property. We will introduce consistency in Sect. 5.2. When using a consortium of proper local pre-experts there is no implication from accepting an implication. An accepted implication may be false in the target domain.

To meet our goal of reducing the number of inquiries to the individual expert in a consortium, the proposed consortial expert from Definition 4.14 is insufficient. We need to diminish the strong requirement from checking all experts for having a counterexample. This leads to the following definition.

Definition 4.15 (Consortial expert). Let $\mathcal{C} = \{p_i\}_{i \in I}$ be a consortium for $\mathcal{X}$ on M. A *consortial expert* is a mapping $p_{\mathcal{C}} \colon \bigcup_{i \in I} \mathrm{Imp}(N_i) \to \bigcup_{i \in I} \mathcal{X}_i \cup \{\top\}$ such that for every $f = (A, B) \in \bigcup_{i \in I} \mathrm{Imp}(N_i)$ there holds:

(1) $\exists p_i \in \mathcal{S}, p_i(f) \neq \top \Rightarrow p_{\mathcal{C}}(f) \neq \top,$
(2) $p_{\mathcal{C}}(f) = X \in \bigcup_{i \in I} \mathcal{X}_i \Rightarrow A \subseteq X \wedge B \nsubseteq X.$

The set $\mathcal{S} \subseteq \mathcal{C}$ is a per inquiry chosen subset of local experts such that $f \in \mathrm{Imp}(N_i)$ for every $p_i \in \mathcal{S}$.

We left the just addressed expert subset vague by intention. In practice, choosing this should be possible in various ways. There is no further restriction then of choosing "qualified" experts, i.e., how the consortial expert is choosing $\mathcal{S}$. One obvious choice would be to consult all local (pre-)experts at once. A more clever strategy would be to consult experts covering the attributes in question having the largest attribute size to cover in general. One may also employ a cost function, which could lead to asking only less expensive experts. While using a consortial expert for exploration, an already accepted implication may be refuted later on in the exploration process. Whenever an inquiry leads to an counterexample which is also an counterexample for an already accepted implication, the set of valid implications needs to be corrected.

So far we provided neither constraints nor constructions about the decision making of a consortium, i.e., the *collaboration*. The most simple case, where $\mathcal{M}$ is a partition of M and all queries are concerned with an element of $\mathcal{M}$, can easily be treated: For every query the expert for the according element of $\mathcal{M}$ either refutes or maybe accepts. Since this case seems artificial we will investigate different approaches of "real" collaboration in the following section.

5   Exploration with Consortial Experts

In general, for exploring a domain using attribute exploration with partial examples one may use instead of the domain expert some (strong) consortial expert. However, there are three possible problems to deal with. First, a query may concern some implication f that is not well-formed for the consortium $\mathcal{C}$ that is used by the consortial expert. Second, if a consortium containing local pre-experts does accept an implication this does not necessarily imply the implication in question to be valid in the domain. Obviously, this also depends heavily on how a consortial expert utilizes a consortium. We deal with related problems in the following subsections. Third, while choosing a subset of $\mathcal{C}$ the consortial expert may have missed a local pre-expert which would have been aware of a counterexample, in contrast to a strong consortial expert.

The first problem cannot be resolved by the consortial expert. When no local (pre-)expert can be consulted for some implication the only choice is to accept f. However, a more suitable response would be a third type of replying like *NULL*. Then, the exploration algorithm could cope with this problem by deferring to other questions. The attribute exploration algorithm with partial examples from [4] but could easily be adapted for this. In turn, the algorithm would only be able to return an interval of closure systems, like in Fig. 1.

For the second problem one needs to repair the set of accepted implications in case a counterexample turns up later in the process. We show a method of doing so in Sect. 5.4. Of course, there is still the possibility that an accepted not valid implication will never be discovered as a consequence of an incapable consortium. This leads the exploration algorithm to return not the target domain but another closure system. How "close" this closure system is to the target domain, in terms of some Jaccard-like measure, is to be investigated in some future work.

The third and last problem can always be dealt with by employing a strong consortial expert. A less exhaustive method could be to incorporate statistical methods for quantifying the number of necessary experts to consult in order to obtain a low margin of error.

5.1   Correcting Falsely Accepted Implications

A major issue while using a consortial expert for exploration is the possibility of wrongly accepting an implication. This can be dealt with on side of the exploration algorithm. While receiving a new counterexample $O \subseteq M$ from the consortial expert the exploration algorithm has also to check if O is a counterexample to an already accepted implication in $\mathcal{L}$. When such an implication $f = (A, B)$ is found, we would need to restrict the conclusion of f to a yet not disproved subset and also add implications with stronger premises that were omitted because f was (wrongly) accepted. In particular, we would replace f in $\mathcal{L}$ by $A \to B \cap C$ and also add implications $A \cup \{m\} \to B$ for $m \in M \setminus (A \cup C)$ to $\mathcal{L}$.

This approach may drastically increase the size of the set of already accepted implications. Unlike the classical exploration algorithm, this modified version would return a very large set of implications instead of the canonical base. One may cope with that by utilizing [4, Algorithm 19] after every event of replacing an implication in $\mathcal{L}$. This algorithm takes a set of implications and returns the canonical base. After this a next query can be computed based on the so far collected set of implications and the already collected counterexamples.

5.2   Consistency

So far we characterized what local pre-experts and consortia are, by their ability to make decisions about queries. In this section we provide ideas for a consistent consortium. We start with resolving a possible conflict for consortial experts.

Definition 5.1 (Consistent experts). Let $\mathcal{C} = \{p_i\}_{i \in I}$ be a consortium for $\mathcal{X}$ on M and let $\check{\mathcal{C}} \subseteq \mathcal{C}$ be the set of local experts in $\mathcal{C}$. We say that $\mathcal{C}$ has *consistent experts* if for $i, j \in I$ with $p_i, p_j \in \check{\mathcal{C}}$ and for all $f \in \mathrm{Imp}(N_i \cap N_j)$ it holds that $p_i(f) = \top \Leftrightarrow p_j(f) = \top$.
   We call $\mathcal{C}$ with consistent experts a *consistent experts consortium*.

This idea from consistent experts does still allow for different local experts to be able to refute an implication with different counterexamples. But whenever one local expert would accept an implication, any other local expert needs to do so as well. Different local (pre-)experts may have access to disjoint sets of counterexamples, by design. Furthermore, local pre-experts may not have the knowledge for all possible counterexamples in their restriction of the target domain. Therefore, accepting an implication by a local pre-expert has no implication itself. Hence, even in a consistent experts consortium it is still possible that some local experts may provide a counterexample while others do not. A stronger notion of consistency would be to forbid that.

Definition 5.2 (Consistent consortium). Let $\mathcal{C} = \{p_i\}_{i \in I}$ be a consortium for $\mathcal{X}$ on M. The consortium $\mathcal{C}$ is *consistent* if for all $i, j \in I$ and for all $f \in \mathrm{Imp}(N_i \cap N_j)$ we have that $p_i(f) = \top \Leftrightarrow p_j(f) = \top$.

Again, in consequence, all local pre-experts are either able to produce some, but not necessarily the same, counterexample for some implication or all do accept. We look into the possibility of combining counterexamples in Sect. 5.4.

5.3   Abilities and Limitations of a Consortium

In this section we exhibit the theoretical abilities and limitations of a consortium for determining the whole target domain of available knowledge. After clarifying some general notions and facts, we state a reconstructability result for consortia based on combinatorial designs.
   Let us, as before, fix a finite (attribute) set M. As is well-known, any set $\mathcal{F} \subseteq \mathrm{Imp}(M)$ of implications constitutes a closure system

$$\mathcal{X}_\mathcal{F} := \{X \in \mathcal{P}(M) \mid \forall f = (A, B) \in \mathcal{F} : A \subseteq X \Rightarrow B \subseteq X\}.$$

Conversely, any closure system $\mathcal{X}$ defines its set $\mathcal{F}_\mathcal{X} \subseteq \mathrm{Imp}(M)$ of valid implications, and we have $\mathcal{X}_{\mathcal{F}_\mathcal{X}} = \mathcal{X}$ and $\mathcal{F}_{\mathcal{X}_\mathcal{F}} = \mathcal{F}$. Now suppose that $\mathcal{S}$ is a class of closure systems $\mathcal{X} \subseteq \mathcal{P}(M)$ on M which contains the target domain. This set $\mathcal{S}$ describes some information on the target domain we may have in advance. Suppose that $\mathcal{M} = \{N_i \mid i \in I\}$ is a consortial domain and we have, for some

$\mathcal{X} \in \mathcal{S}$, a set of local experts $p_i \colon \mathrm{Imp}(N_i) \to \mathcal{X}_{N_i} \cup \{\top\}$ on $N_i \in \mathcal{M}$, so that in particular, $p_i(f) = \top$ if and only if f is valid in $\mathcal{X}$. Then we consider the set

$$\mathcal{F}_{\mathcal{M}} := \{f \in \textstyle\bigcup_{i \in I} \mathrm{Imp}(N_i) \mid f \text{ is valid}\} \subseteq \mathcal{F}_{\mathcal{X}},$$

i.e., the set of all well-formed valid implications, and let $\mathcal{X}_{\mathcal{M}} := \mathcal{X}_{\mathcal{F}_{\mathcal{M}}}$, which is the closure system reconstructible by the consortium. Clearly, $\mathcal{X}_{\mathcal{M}} \supseteq \mathcal{X}$, and from the preceding discussion we easily deduce the following result.

Proposition 5.3 (Ability of a consortium). *The consortial domain $\mathcal{M}$, together with local experts $p_i \colon \mathrm{Imp}(N_i) \to \mathcal{X}_{N_i} \cup \{\top\}$ for $N_i \in \mathcal{M}$, is able to reconstruct the target domain $\mathcal{X}$ within a class $\mathcal{S}$ of closure systems on M if and only if $\mathcal{Y}_{\mathcal{M}} = \mathcal{X}_{\mathcal{M}}$ implies $\mathcal{Y} = \mathcal{X}$, for every $\mathcal{Y} \in \mathcal{S}$.*

Example 5.4. We illustrate these notions with two simple extreme cases.

1. Suppose that $\mathcal{X} = \{M\}$, then every implication is valid, i.e., $\mathcal{F}_{\mathcal{X}} = \mathrm{Imp}(M)$. Since every consortial domain $\mathcal{M} = \{N_i \mid i \in I\}$ has the covering property $\bigcup_{i \in I} N_i = M$, it follows that $\mathcal{X}_{\mathcal{M}} = \mathcal{X}$. Hence, if $\mathcal{Y}_{\mathcal{M}} = \mathcal{X}_{\mathcal{M}}$, then $\mathcal{Y}_{\mathcal{M}} = \{M\}$, so that $\mathcal{Y} = \{M\} = \mathcal{X}$, i.e., the consortium is always able to reconstruct $\mathcal{X}$ in the class of all closure systems on M.
2. Consider the case $\mathcal{X} = \mathcal{P}(M)$ and suppose that $\mathcal{M} = \{N_i \mid i \in I\}$ is a proper consortial domain. Then for any $m \in M$ we have $M \setminus \{m\} \to \{m\} \notin \bigcup_{i \in I} \mathrm{Imp}(N_i)$, whence $\mathcal{X}_{\mathcal{M}} = \mathcal{Y}_{\mathcal{M}}$ for $\mathcal{Y} = \mathcal{P}(M) \setminus \{M \setminus \{m\}\} \neq \mathcal{X}$. Thus no proper consortium is capable of reconstructing the target domain.

Let us define for a set of implications $\mathcal{F} \subseteq \mathrm{Imp}(M)$ the *premise complexity* to be $c(\mathcal{F}) := \max\{|A| \mid f = (A, B) \in \mathcal{F}\}$ if $\mathcal{F} \neq \varnothing$ and $c(\varnothing) := -1$. Also, we associate to a closure system $\mathcal{X} \subseteq \mathcal{P}(M)$ on M its *premise complexity* by $c(\mathcal{X}) := \min\{c(\mathcal{F}) \mid \mathcal{X}_{\mathcal{F}} = \mathcal{X}\}$, which equals the premise complexity of its canonical base.

Example 5.5. For the extreme closure systems we have $c(\mathcal{P}(M)) = -1$ and $c(\{M\}) = 0$. Considering the closure system $\mathcal{X}_k := \{X \in \mathcal{P}(M) \mid |X| \leq k\} \cup \{M\}$ we see that $c(\mathcal{X}_k) = k + 1$.

Denote by $\mathcal{S}_k$ the class of all closure systems up to premise complexity k.

Theorem 5.6 (Reconstructability in bounded premise complexity). *A consortium of local experts on the consortial domain $\mathcal{M}$ is able to reconstruct a target domain within the class $\mathcal{S}_k$ if and only if every subset $O \subseteq M$ of size $k+1$ is contained in some $N \in \mathcal{M}$.*

Proof. First suppose that each subset $O \subseteq M$ of size $k+1$ is contained in some $N \in \mathcal{M}$. We claim that $\mathcal{X}_{\mathcal{M}} = \mathcal{X}$ for every closure system $\mathcal{X} \in \mathcal{S}_k$, whence every target domain is reconstructible within $\mathcal{S}_k$. Let $\mathcal{X} \in \mathcal{S}_k$, then there is a set $\mathcal{F}$ of implications with premise complexity $c(\mathcal{F}) \leq k$ such that $\mathcal{X} = \mathcal{X}_{\mathcal{F}}$. We may assume that each implication $f \in \mathcal{F}$ is of the form $f = (A, \{b\})$. By assumption there holds $\mathcal{F} \subseteq \bigcup_{N \in \mathcal{M}} \mathrm{Imp}(N)$, so that $\mathcal{F} \subseteq \mathcal{F}_{\mathcal{X}} \cap \bigcup_{N \in \mathcal{M}} \mathrm{Imp}(N) = \mathcal{F}_{\mathcal{M}} \subseteq \mathcal{F}_{\mathcal{X}}$. This implies $\mathcal{X}_{\mathcal{F}} = \mathcal{X}_{\mathcal{F}_{\mathcal{M}}}$, i.e., $\mathcal{X}_{\mathcal{M}} = \mathcal{X}$, as desired.

Conversely, suppose there exists a subset $O \subseteq M$ of size $k+1$ not contained in any $N \in \mathcal{M}$. Choose some $b \in O$, let $A := O \setminus \{b\}$, so that $|A| = k$, and consider the implication $f := (A, \{b\})$. Then we have $f \notin \bigcup_{N \in \mathcal{M}} \mathrm{Imp}(N)$. Now letting $\mathcal{X} := \mathcal{P}(M)$ and $\mathcal{Y} := \mathcal{X}_{\{f\}}$ we then have distinct $\mathcal{X}, \mathcal{Y} \in \mathcal{S}_k$ with $\mathcal{X}_{\mathcal{M}} = \mathcal{P}(M) = \mathcal{Y}_{\mathcal{M}}$, showing that $\mathcal{X}$ cannot be reconstructed within $\mathcal{S}_k$. $\square$

Suppose that $|M| = m$. Recall (cf. [8, Sect. 2.5]) that a *Steiner system* $S(t, n, m)$ is a collection $\{N_i \mid i \in I\} \subseteq \mathcal{P}(M)$ of n-element subsets $N_i \subseteq M$ such that every t-element subset of M is contained in exactly one subset N_i. In light of Theorem 5.6 it is clear that the Steiner systems $S(k+1, m, n)$ are the minimal consortial domains that are able to reconstruct target domains within the class $\mathcal{S}_k$ of bounded premise complexity k.

5.4   Extensions

Combining Counterexamples. A lack of our consortium setting so far is the inability to recognize similar counterexamples. Combining counterexamples is a powerful idea that lifts the consortium above the knowledge of the "sum" of knowledge of the local (pre-)experts.

For this we need to augment a consortium by a background ontology of counterexamples. The most simple approach would be to identify two counterexamples from two different local (pre-)experts by matching the names of the counterexamples, which the experts would need to provide as well. In basic terms of FCA, while providing counterexamples the consortial expert needs to know if the counterexamples provided by the local (pre-)experts, restricted to their attribute sets, are of the same counterexample in the domain. We motivate this extension by an example. Given we want to explore some domain about animals with the attribute set being $M = \{$mammal, does not lay eggs, is not poisonous$\}$ using a set of two local (pre-)experts with $N_1 = \{$mammal, does not lay eggs$\}$ and $N_2 = \{$mammal, is not poisonous$\}$. Only expert p_1 can be consulted for the validity of $\{$mammal$\} \rightarrow \{$does not lay eggs$\}$. Of course, p_1 refutes this implication by providing the set $\{$mammal$\}$, which he could name for example *platypus*[3]. While exploring, the next query could be $\{$mammal$\} \rightarrow \{$is not poisonous$\}$. Note that this is not answered by the counterexample of p_1 since $\{$is not poisonous$\}$ is no subset of N_1. The local (pre-)expert p_2 refutes this of course as well, by providing the counterexample $\{$mammal$\}$ and naming this counterexample also platypus. Instead of collecting two different counterexamples we are now able to combine those two and say $\{$mammal$\}$ is not just an element of $\mathcal{X}_1$ and $\mathcal{X}_2$ but as well an element of $\mathcal{X}$. In turn, the set of counterexamples the exploration algorithm is using contains now a more powerful counterexample than any local expert in the consortium could have provided. There are various ways to implement this combining of counterexamples. For example, after acquiring a

---

[3] a semiaquatic egg-laying mammal endemic to eastern Australia.

counterexample for an implication from some expert one may ask all experts if they are aware of this counterexample name and if they could contribute further attributes from their local attribute set. To investigate efficient strategies to do that is referred to future work.

Coping with Wrong Counterexamples. Another desirable ability for a real world consortium would being able to handle wrong counterexamples, or more generally, having a measure that reflects the trust a consortial expert has in counterexamples of particular local (pre-)experts. Our setting for a consortium is not capable of doing this. In fact, the consortium cannot refute an implication using a wrong counterexample by design, since every $\mathcal{X}_i$ is a restriction of the target domain. Hence, all counterexamples provided by a local (pre-)expert are "true". In order to allow for a consortium to provide wrong counterexamples, one has to detach the closure system of some expert p_i from the target domain $\mathcal{X}$. This would also extend the possibilities of treating counterexamples by the consortial expert. Resolution strategies from simple majority voting up to minimum expert trust or confidence could be used.

6   Conclusion and Outlook

In this paper we gave a first characterization of how to distribute the rôle of an domain expert for attribute exploration onto a consortium of local (pre-)experts. Besides practically using this method this result may be applied to various other tasks in the realm of FCA. It is obvious that the shown approach can easily be adapted for object exploration, the dual of attribute exploration. Hence, having object and attribute exploration through a consortium, we provided the necessary tools such that collaborative concept exploration (CCE) is at reach. Since CCE relies on both kinds of exploration in an alternating manner, the logical next step is to investigate what can be explored using a consortium. In addition we showed preliminary results on how to evaluate a consortium, how to shape it, i.e., how to choose a consortium from a potentially bigger set of experts, how to treat mistakenly accepted implications and how to increase the consistency.

Further research on this could focus on formalizing the depicted extensions from Sect. 5.4, where the task of modifying the consortium in order to encounter and compute wrong counterexamples seems as inevitable as it is hard to do. An easier extension that increases the ability of exploring a domain seems to be a "clever" combining of counterexamples.

Acknowledgments. The authors would like to thank Daniel Borchmann and Maximilian Marx for various inspiring discussions on the topic of consortia while starting this project. In particular, the former suggested the name *consortium* and always is the best critic one can imagine. Furthermore, we are grateful for various challenging discussions with Sergei Obiedkov, including ideas for coping with wrongly accepted implications.

References

1. Ågotnes, T., Wáng, Y.N.: Resolving distributed knowledge. Artif. Intell. **252**, 1–21 (2017)
2. Burmeister, P., Holzer, R.: Treating incomplete knowledge in formal concept analysis. In: Ganter, B., Stumme, G., Wille, R. (eds.) Formal Concept Analysis. LNCS (LNAI), vol. 3626, pp. 114–126. Springer, Heidelberg (2005). https://doi.org/10.1007/11528784_6
3. Ganter, B.: Attribute exploration with background knowledge. Theor. Comput. Sci. **217**(2), 215–233 (1999)
4. Ganter, B., Obiedkov, S.A.: Conceptual Exploration, pp. 1–315. Springer, Heidelberg (2016). https://doi.org/10.1007/978-3-662-49291-8
5. Ganter, B., Wille, R.: Formal Concept Analysis: Mathematical Foundations. Springer, Heidelberg (1999). https://doi.org/10.1007/978-3-642-59830-2. pp. x+284
6. Holzer, R.: Knowledge acquisition under incomplete knowledge using methods from formal concept analysis: part II. Fundam. Inform. **63**(1), 41–63 (2004)
7. Jäger, G., Marti, M.: A canonical model construction for intuitionistic distributed knowledge. In: Advances in Modal Logic, pp. 420–434. College Publications (2016)
8. MacWilliams, F.J., Sloane, N.J.A.: The Theory of Error-Correcting Codes, p. 369. North-Holland Publishing Co., Amsterdam (1977). pp. xv+369
9. Martin, P., Eklund, P.W.: Large-scale cooperatively-built KBs. In: Delugach, H.S., Stumme, G. (eds.) ICCS-ConceptStruct 2001. LNCS (LNAI), vol. 2120, pp. 231–244. Springer, Heidelberg (2001). https://doi.org/10.1007/3-540-44583-8_17
10. Obiedkov, S.: Modal logic for evaluating formulas in incomplete contexts. In: Priss, U., Corbett, D., Angelova, G. (eds.) ICCS-ConceptStruct 2002. LNCS (LNAI), vol. 2393, pp. 314–325. Springer, Heidelberg (2002). https://doi.org/10.1007/3-540-45483-7_24
11. Obiedkov, S., Romashkin, N.: Collaborative conceptual exploration as a tool for crowdsourcing domain ontologies. In: RuZA 2015. CEUR Workshop Proceedings, CEUR-WS.org, vol. 1552, pp. 58–70 (2015)
12. Rudolph, S.: Relational exploration: combining description logics and formal concept analysis for knowledge specification. Ph.D. thesis. Technische Universität Dresden, Germany (2006)
13. Stange, D., Nürnberger, A., Heyn, H.: Collaborative knowledge acquisition and exploration in technology search. In: 18. GeNeMe-Workshop 2015. Technische Universität Dresden, pp. 243–249 (2015)
14. Stumme, G.: Concept exploration: knowledge acquisition in conceptual knowledge systems. Ph.D. thesis. Darmstadt University of Technology (1997)
15. Tang, M.T., Toussaint, Y.: A collaborative approach for FCA-based knowledge extraction. In: Concept Lattices and Their Applications CLA. CEUR Workshop Proceedings, CEUR-WS.org, vol. 1062, pp. 281–286 (2013)
16. Vrandečić, D., Krötzsch, M.: Wikidata: a free collaborative knowledgebase. Commun. ACM **57**(10), 78–85 (2014)

Graph Visualization

A Visual Analytics Technique for Exploring Gene Expression in the Developing Mouse Embryo

Simon Andrews[1(✉)] and Kenneth McLeod[2]

[1] Conceptual Structures Research Group, Communication and Computing Research Centre and The Department of Computing, Faculty of Arts, Computing, Engineering and Sciences, Sheffield Hallam University, Sheffield, UK
`s.andrews@shu.ac.uk`
[2] School of Mathematical and Computer Sciences, Heriot-Watt University, Edinburgh, UK
`kenneth.mcleod@hw.ac.uk`

Abstract. This paper describes a novel visual analytics technique for exploring gene expression in the developing mouse embryo. The majority of existing techniques either visualise a single gene profile or a single tissue profile, whereas the technique presented here combines both - visualising the genes expressed in each tissue in a group of tissues (the components of the developing heart, for example). The technique is presented using data, provided by the Edinburgh Mouse Atlas Project, of gene expression assays conducted on tissues of the developing mouse embryo and a corresponding hierarchical graph of tissues defining the mouse anatomy. By specifying a particular tissue, such as the heart, and a particular stage of development, a Formal Context is computed making use of the hierarchical mouse anatomy so that the resulting Formal Concept Lattice visualises the components of the specified tissue and the genes expressed in each component. An algorithm is presented that defines the computation the Formal Context. Examples of resulting lattices are given to illustrate the technique and show how it can provide useful information to researchers of gene expression and embryo development.

1 Introduction

Understanding the role of genes in the development of an embryo is a major scientific endeavour. Key advances such as the mapping of the human genome[1], the identification of specific genes responsible for genetic diseases [1] and the latest gene splicing technologies, such as the CRISPR-cas9 system [2], have led to new methods and treatments for the detection, prevention and correction of many genetic disorders [3]. However, there is still much to be done to gain a

---

[1] https://www.genome.gov/.

© Springer International Publishing AG, part of Springer Nature 2018
P. Chapman et al. (Eds.): ICCS 2018, LNAI 10872, pp. 137–151, 2018.
https://doi.org/10.1007/978-3-319-91379-7_11

complete picture of how the many thousands of genes present in complex organisms are responsible for the construction, differentiation and organisation of the many different cells that constitute the organism. By studying the components of a developing embryo, in terms of finding out which genes are responsible for constructing which systems, organs and tissues therein, a fuller picture is emerging.

One major effort in this regard is the e-Mouse Atlas Project [4,5] where in situ gene expression experiments for the embryonic mouse are aggregated and published. Data is being collected that identifies which, of over 6000 genes, are responsible for the construction of over 4000 different tissues in the developing mouse embryo.

This paper presents a new visual analytics technique to explore this data, allowing the visualisation of the genes expressed in a specified mouse embryo tissue and all of its component parts (such as the heart and its components, comprising atria, ventricles, valves, etc.). The visualisation also enables gene co-expression to be seen, where groups of genes are expressed together in the same tissue.

For this purpose Formal Concept Analysis (FCA) [6] is used to create the visualisation in the form of a Formal Concept Lattice. Whereas the majority of existing techniques tend to focus on either a single gene profile across different tissues, or single tissue profile of genes expressed therein, the Formal Concept Lattice of gene expression combines both in a single diagram (for those not familiar with FCA, these are good introductory texts: [7,8]).

The structure of the rest of this paper is as follows: Sect. 2 describes the e-Mouse Atlas Project (EMAP), Sect. 3 shows how the EMAP data and a corresponding definition of the mouse anatomy can be used to create a Formal Context of gene expression, Sect. 4 evaluates the results by means of a number of Formal Concept Lattices of gene expression, Sect. 5 is a review of existing techniques and other related work and Sect. 6 draws conclusions from the work and suggests further work to be done.

2   The e-Mouse Atlas Project

The e-Mouse Atlas Project (EMAP) provides researchers with two main resources: the EMA Mouse Anatomy Atlas and the EMAGE Gene Expression Database [9].

The EMA Mouse Anatomy Atlas uses embryological mouse models to provide a digital atlas of mouse development. It is based on the definitive books of mouse embryonic development by Theiler [10] and Kaufman [11] yet extends these studies by creating a series of three dimensional computer models of mouse embryos at successive stages of development with defined anatomical domains linked by a stage-by-stage ontology of anatomical tissue names.

The stages of mouse embryonic development in the EMA Mouse Anatomy Atlas are based on those defined by Theiler [10]. Theiler defined 26 stages of embryonic development, from Theiler Stage 1: *the one-cell egg* to Theiler Stage

26: *Long Whiskers.* In EMAP, each Theiler Stage is represented in an anatomical ontology (from Kaufman [11]) and a corresponding three dimensional computer model. Figure 1 shows part of the anatomy ontology at Theiler Stage 13, where *cardiovascular system* and *heart* have been expanded to show their component parts. Each anatomical term is annotated with a unique ID (EMAPA number) and the Theiler Stages in which it is present. Note that, although Fig. 1 resembles a taxonomy, it is actually an ontology: a parent tissue can have several child tissues but also a child tissue can have more than one parent.

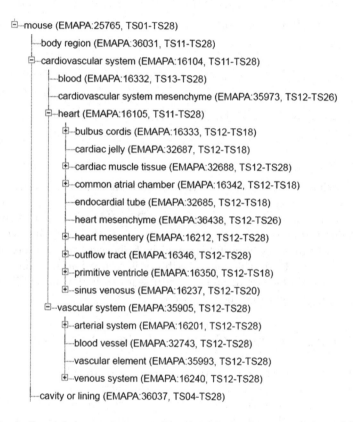

Fig. 1. Part of the anatomy ontology for Theiler Stage 13. Source: [5].

EMAGE is a database of *in situ* gene expression data in the mouse embryo and an accompanying suite of tools to search and analyse the data. The majority of the data is based on whole-mount (whole embryo) and embryo section assays. The embryo or section is stained-up for a particular gene of interest to show where the gene is being expressed. The scientist carrying out the assay has to determine in which tissues the gene is being expressed and how strongly it is expressed by the level of staining. The strength of expression in a tissue is denoted as one of a number of possible values: *detected, not detected, possible, strong, moderate* or *weak*. Table 1 is a representation of EMAGE data, with values for Theiler Stage, *emageID* (a unique assay number), *gene, strength, tissueName* and *emapaId* (a unique tissue ID). Note that several tissues can be examined in the same assay and, that for the purposes of data analysis, all components of the embryo are called tissues, including organs, systems and even the mouse itself.

Table 1. A representation of EMAGE data.

Stage	emageID	Gene	Strength	tissueName	emapaId
17	503	Mfng	Detected	cranial ganglion	16659
17	503	Mfng	Detected	hindlimb ridge	16593
17	504	Actc1	Strong	dorsal aorta	16204
17	504	Actc1	Strong	branchial arch	16117
15	505	Zeb1	Detected	branchial arch	16117
15	505	Zeb1	Detected	cranial ganglion	16659
15	505	Zeb1	Strong	hindlimb ridge	16593

Figure 2 shows an example of a whole-mount assay and the resulting spatial annotation for gene Gpc3 at Theiler Stage 15: The results of the whole-mount assay (a) are mapped onto a corresponding 2D section of the model (b) using a colour-coding to indicate the strength of expression. However, this type of visualisation is not particularly useful if we wish to show *all* the genes expressed in a group of tissues. To do this, one possibility is to use Formal Concept Analysis, taking genes to be formal objects and tissues to be formal attributes. This technique is described in the next Section.

3   Creating a Formal Context of Gene Expression

A formal context of gene expression can be created from EMAGE data taking genes to be formal objects and tissues to be formal attributes. If gene g is detected in tissue t, for example, the relation (g, t) is added to the context. Of course, this approach loses the detail of strength of expression and is useful only if we are interested in whether a gene has been detected or not. In which case, the strengths *weak, moderated* and *strong* can all be taken to imply *detected*.

Taking the EMAGE data in Table 1 as an example, a corresponding formal context is shown in Fig. 3(a) and the resulting concept lattice in Fig. 3(b)[2]. Thus gene *Actc1* is expressed in the *dorsal aorta* and the *branchial arch*, *Zeb1* is expressed in the *branchial arch*, the *hindlimb ridge* and the *cranial ganglion*, and *Mfng* is expressed in the *hindlimb ridge* and the *cranial ganglion*.

---

[2] The lattices presented in this paper were all produced using Concept Explorer [12] available at https://sourceforge.net/projects/conexp/.

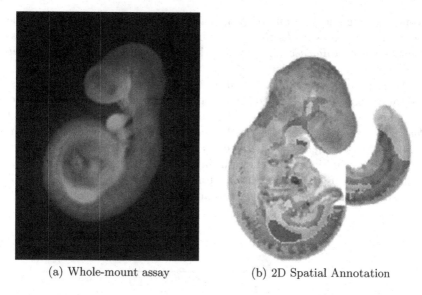

(a) Whole-mount assay   (b) 2D Spatial Annotation

Fig. 2. Whole mount assay and spatial annotation of gene Gpc3 at TS15. Name: EMAGE 3837. Source: [9].

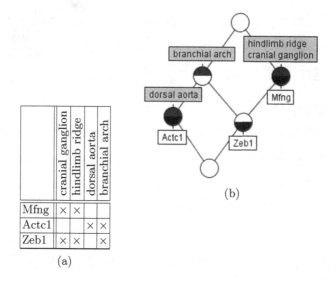

(a)

(b)

Fig. 3. Formal context and resulting lattice derived from the EMAGE data extract in Table 1.

3.1   Inferred Gene Expression Results

The hierarchical nature of the mouse anatomy ontology means that additional results can be inferred from an assay. If a gene is detected in a tissue then, by definition, it is also expressed in that tissue's parent tissue, and so on up the anatomy. Taking the part anatomy shown in Fig. 1 as an example, if a gene is detected in the *bulbus cordis* then it can be inferred that the gene is expressed in the *heart*, the *cardiovascular system* and the *mouse*. This is known as *positive propagation* of gene expression.

This positive propagation is useful to FCA in producing a gene expression visualisation that corresponds to the hierarchy in the mouse anatomy ontology. For example, in a number of assays let us say that *gene1* was detected in the *bulbus cordis*, *gene2* was detected in the *heart*, *gene3* was detected in the *cardio-vascular system* and *gene4* was detected in the *mouse*. The corresponding formal context is shown in Fig. 4a and the resulting lattice in Fig. 4b. The lattice clearly visualises the hierarchy of the mouse anatomy ontology thus providing a familiar and sensible structure for researchers to analyse.

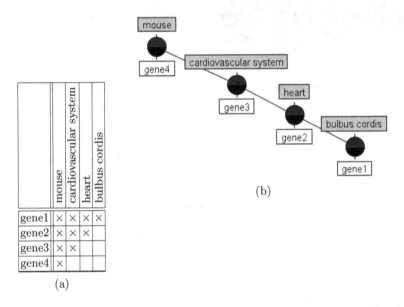

Fig. 4. Formal context and resulting lattice demonstrating positive propagation of gene expression.

3.2   An Algorithm for Creating Formal Contexts of Gene Expression from EMAGE Data

Although it is possible to create a formal context from all of the data the EMAGE database, the resulting concept lattice would be far to large to analyse. Instead, what is proposed here is to create a formal context of gene expression for a particular tissue of interest (such as an organ or system) and all of its component parts at a given Theiler Stage. The algorithm, *FindDetectedGenesInTissue*, presented below, automates the construction of such formal contexts.

The algorithm is invoked by passing it the tissue and Theiler Stage of interest and an initially empty set of tissues called *family_line*. This set of tissues is used to store (in line 2 of the algorithm) an ancestral line of tissues from the initial tissue of interest, to a child of that tissue, a grandchild and so on. This family line of tissues provides the means of carrying out positive propagation of gene expression.

In line 3 of the algorithm, the EMAGE database is searched to find results involving the tissue and Theiler Stage of interest. For each such result, if the strength of the result is *detected, strong, moderate* or *weak*, then, in line 6 and 7, for each tissue in the current family line, a relation between each tissue and the gene in the result is added to the formal context being created, thus carrying out positive propagation of the gene's expression.

The algorithm is then recursive, passing to itself successively each child of the current tissue along with the original Theiler Stage and the current family line of tissues. The mouse anatomy for each Theiler Stage is stored as a data set of $(childtissue, parenttissue)$ pairs. Thus is it simply a case of searching for each instance of a parent tissue to enumerate its children.

FindDetectedGenesInTissue$(tissue, tstage, family\_line)$

1 **begin**
2 $\quad$ $family\_line \leftarrow family\_line \cup \{tissue\}$
3 $\quad$ **foreach** $result$ in $EMAGEdatabase$ **do**
4 $\quad\quad$ **if** $result.tissueName = tissue$ **and** $result.stage = tstage$ **and**
$\quad\quad$ $(result.strength = \text{"detected"}$ **or** $result.strength = \text{"strong"}$ **or**
$\quad\quad$ $result.strength = \text{"moderate"}$ **or** $result.strength = \text{"weak"})$ **then**
5 $\quad\quad\quad$ **remark** positive propagation of gene expression:
6 $\quad\quad\quad$ **foreach** $Tissue$ in $family\_line$ **do**
7 $\quad\quad\quad\quad$ $context \leftarrow context \cup \{(result.gene, Tissue)\}$

8 $\quad$ **foreach** $child$ of $tissue$ in $MouseAnatomy(tstage)$ **do**
9 $\quad\quad$ FindDetectedGenesInTissue $(child, tstage, family\_line)$

10 **end**

---

4   Concept Lattices of Gene Expression

The *FindDetectedGenesInTissue* algorithm was implemented in C++ and a series of gene expression analyses were carried out. The first analysis is of the *eye* and its components at Theiler Stage 16 and the second analysis of the development of the *heart atrium* over a number of Theiler Stages.

Figure 5 is a lattice showing gene expression in the *eye* at Theiler Stage 16. The genes expressed in the eye and its various components are clearly seen. In some cases, such as the *optic cup*, there appears quite a large number of co-expressed genes. This may indicate that the tissue is composed of many different proteins or may be an example of several genes working together to produce a single protein - sometimes one or more 'enabler' genes are required for the expression of the protein making gene [13]: the products of two genes, for example, (called *transcription factors*) work together to influence the expression of another gene and therefore the amount of protein product of that gene.

Other, more specialised tissues, such as the *optic cup intraretinal space*, show the expression of a single gene, which may indicate that the tissue is predominantly composed of a single protein. There are also interesting cases of genes, such as *Cdh2*, that are expressed in two or more unrelated tissues (tissues that are not in the same branch in the anatomy ontology), in this case the *optic cup* and the *lens pit*. This may indicate that, although not directly related, the tissues share some similarity in function or structure. It is also interesting to look at gene expression in tissues that clearly, from their names, have some commonality and often share the same parent tissue, such as the *optic cup outer layer* and the *optic cup inner layer*. Here we can see that gene *Wnt2b* is expressed in both layers, but also that each layer has genes specific to it. The shared gene may be responsible for a common function or structure whilst the specific genes may be responsible for the structural or functional differences between the tissues.

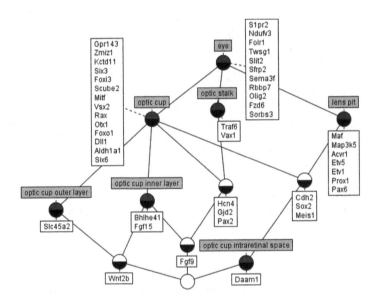

Fig. 5. Lattice of gene expression in the *eye* at Theiler Stage 16.

Figures 6 and 7 show lattices of gene expression in the *heart atrium* across four Theiler Stages: 14 (Fig. 6(a)), 15 (Fig. 6(b)), 16 (Fig. 7(a)) and 17 (Fig. 7(b)). The lattices clearly show changes in gene expression during the development of the *heart atrium* over these Stages. In the earliest Stage, there are only four components and three genes. The atrium at this stage consists primarily of *cardiac muscle* being constructed by two genes, *Vcam1* and *Myl9*. The other

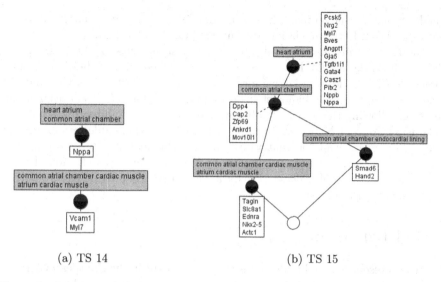

(a) TS 14   (b) TS 15

Fig. 6. Lattices of gene expression in the *heart atrium* across Theiler Stages 14 and 15.

gene, *Nppa*, continues to be expressed in the *heart atrium* through Stages 15 and 16. Its early presence and the subsequent proliferation of genes expressed with it in the *heart atrium* in TS 15 may suggest that it has a role as controlling gene or *core transcription factor* [14,15], responsible for the activation of other genes during the *heart atrium* development. TS 15 introduces the *common atrial chamber endocardial lining* with two genes being expressed therein. The two *cardiac muscle* genes, *Vcam1* and *Myl7*, in TS 14 are now replaced in TS 15 by five new genes. It is interesting to note that gene *Myl7* seems to have moved from the *cardiac muscle* up to the *heart*. It must be noted, however, that the EMAGE database is not yet complete and this may be a case where *Myl7* has not been assayed in the *cardiac muscle* at TS 15. If, however, it has and was found not to be detected, then this would represent an unusual and interesting finding. In TS 16 the *common atrial chamber* divides into two parts, *left* and *right*, and the gene *Pitx2*, having appeared in the *heart atrium* in TS 14, now moves down to the *common atrial chamber left part*, adding to the suggestion that it was activated in TS 15 (perhaps by *Nppa*) and is now being expressed in the specific tissue for which it is responsible. In the course of development, an early, undifferentiated tissue will have many genes activated within it that are then responsible for the development of the subsequent components of that tissue. The *cardiac muscles* fail to appear in the TS 16 lattice but then reappear in the TS 17 lattice. Clearly they do not physically disappear and then reappear, so again this may be a case where the data is incomplete. It is also striking by how much gene expression changes over a small time-frame of development and in the same tissues. In fact, very few of the genes in the *heart atrium* lattices appear more than once and often the genes expressed in a particular tissue at one Theiler Stage are replaced by a completely different set of genes in the next

Theiler Stage. At the moment, how much of this is due to rapid changes in tissue development and how much is due to missing data, is not known.

The results were also evaluated by a biologist working on the EMAP. The biologist liked the smaller lattices (such as those presented here) but found larger lattices hard to read (the lattice for the brain at TS 18 has 25 nodes and 137 genes, for example). The biologist suggested that a 'dynamic' visualisation might be useful, allowing the user to "walk through the nodes". Whilst this is not possible using Concept Explorer, an implementation of expandable formal concept trees is, however, available that would provide this feature [16]. Being able to compare gene expression over a number of Theiler Stages was also found to be useful, but the incompleteness of the data set is currently proving to be a barrier to producing a full picture.

5   Related Work

The work presented here builds on research carried out in the European CUBIST Project [17], where novel FCA-based analytics were being developed for a number of use-cases, one of which was the EMAP. One strand of CUBIST developed a technique for identifying large gene co-expressions, typically at the system level (e.g., in the skeletal system) [18]. Thus, for example, the work was able to identify all the genes involved with the construction of the skeletal system. Because of the large amount of missing data in EMAGE (simply due to the large number of genes and tissues involved) this technique employed an element of FCA 'fault tolerance' [19] to predict possible gene expressions at the system level.

Other related work in CUBIST included an investigation of alternative visualisations to the concept lattice [20]. Following this investigation, [21] explored the use of sunburst diagrams to visualise individual gene profiles. Another strand of CUBIST involved the development of an RDF version of the EMAGE database and mouse anatomy ontology, along with novel gene expressions queries made possible via the SPARQL query language [22]. The prototype query tool developed comes closest to the work presented here, in that SPARQL queries may be possible that perform the same task as *FindDetectedGenesInTissue*. In any case, the mouse anatomy ontology used in the work has since been superseded by the current one in EMAP, which would require a re-construction of the RDF model and data to replicate the work. Useful results were obtained, however, in using SPARQL queries to find contradictory results in the EMAGE data, e.g., where one result for a particular gene in a particular tissue at a particular Theiler Stage says *detected* but another result with the same gene, same tissue (or a tissue that the first tissue is a part of) and same Theiler Stage says *not detected*.

FCA appears to be attractive in the study of gene co-expression because formal concepts are natural representations of maximal groups of co-expressed genes. For example, in [23] FCA was used to extract groups of genes with similar expressions profiles from data of the fungus *Laccaria bicolor* and in [24] human SAGE data provided the example from which clusters of genes with similar properties are visualised. In both approaches the complexity, in terms of

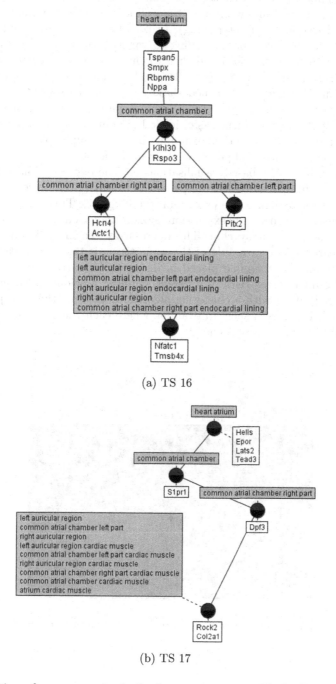

(a) TS 16

(b) TS 17

Fig. 7. Lattices of gene expression in the *heart atrium* across Theiler Stages 16 and 17.

the large number of formal concepts present in the raw data, was managed by specifying a concept's minimum size (the well known idea of minimum support in FCA and frequent itemset mining). In [24], tools were developed to query the set of extracted concepts according to various criteria (e.g., presence of a keyword in a gene description) and then to cluster concepts according to similarity, in terms of the attributes (samples) and objects (genes above a threshold of expression) in them. They called these clusters *quasi-synexpression-groups* (QSGs). By contrast, in [25,26], ranges of a measure of gene concentration were used as attributes and the genes as objects. Individual concepts that satisfied a specified minimum size were then examined by, for example, plotting the actual measures of concentration of genes together in a line plot.

Of course, non-FCA based visualisations of gene expression have also been used by researchers. A notable example is the heatmap, often used to visualise the results of micro-array gene expression profiling [27]. This is a technique for simultaneously measuring the expression levels of thousands of genes for a single sample on one micro-array chip. The micro-array technique is often used in clinical research where a sample of the same tissue is taken from each patient in the study and each sample placed in its own micro-array chip. In order to visualize the micro-array data of different samples, a colour-coded heatmap is generally used, along with a clustering algorithm, allowing the gene profiles to be compared. Figure 8 is such an example where each row is a cancer tumour sample and each column is a gene [28].

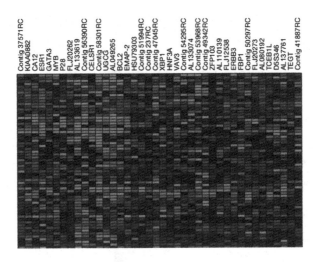

Fig. 8. Example of a heat-map of gene expression in cancer tumours. Source: [28].

6 Conclusions and Further Work

This paper has demonstrated that FCA can provide a useful and novel visualisation of gene expression that combines gene and tissue profiles (i.e., allows both genes and tissues to be displayed together in a single graph. An algorithm called *FindDetectedGenesInTissue* has been developed that incorporates a defined mouse anatomy taxonomy to interrogate the EMAGE database of gene expression for all components of a specified tissue of interest. The algorithm also incorporates positive gene expression propagation to produce a natural hierarchical visualisation of gene expression based on the defined mouse anatomy ontology.

It has been less easy to demonstrate the full potential of comparing gene expression over a series of Theiler Stages of embryo development, due to the large quantify of missing data. The database is far from complete, with many assays still to be conducted, but once the database is complete, analyses across Theiler Stages should be more fruitful. Indeed it may be useful to develop a sort of 'Theiler Stage algebra' whereby, for a specific tissue of interest, one Theiler Stage (formal context) can be subtracted from another, for example, to determine the expressions present in the first Stage but not the second. Alternatively, the intersection of two or more Stages could be carried out to determine the expressions that remain constant over that time.

Although researchers are usually interested in where genes are expressed, they are sometimes interested in finding out where genes are *not* expressed. The EMAGE database already records these *not detected* results and it would be a simple task to modify the algorithm and software to carry out the query. However, the incorporation of positive gene expression propagation would need to be replaced by its corollary, *negative propagation*. In this case, if a gene is not detected in a particular tissue, then it can be inferred that it is also not expressed in all the components of that tissue. The algorithm would thus have to be modified accordingly.

Given a complete EMAGE database, a tool can thus be envisaged that includes the *not detected* query along with a simple set of Theiler Stage algebra and a simple user interface that would provide the researcher a useful technique for gene expression analysis of mouse embryo development.

Acknowledgement. This work was part-funded by the European Commission's 7th Framework Programme of ICT, under topic 4.3: Intelligent Information Management, Grant Agreement No. 257403.

References

1. Hamosh, A., Scott, A.F., Amberger, J.S., Bocchini, C.A., McKusick, V.A.: Online Mendelian Inheritance in Man (OMIM), a knowledgebase of human genes and genetic disorders. Nucleic Acids Res. **33**, 514–517 (2005)
2. Ran, F.A., Hsu, P.D., Wright, J., Agarwala, V., Scott, D.A., Zhang, F.: Genome engineering using the CRISPR-Cas9 system. Nat. Protoc. **8**(11), 2281 (2013)

3. Milunsky, A., Milunsky, J.M.: Genetic Disorders and the Fetus: Diagnosis, Prevention, and Treatment. Wiley, New York (2015)
4. Richardson, L., Venkataraman, S., Stevenson, P., Yang, Y., Moss, J., Graham, L., Burton, N., Hill, B., Rao, J., Baldock, R.A., et al.: EMAGE mouse embryo spatial gene expression database: 2014 update. Nucleic Acids Res. **42**(D1), D835–D844 (2013)
5. EMAP eMouse Atlas Project. http://www.emouseatlas.org. Accessed Dec 2017
6. Ganter, B., Stumme, G., Wille, R. (eds.): Formal Concept Analysis: Foundations and Applications. LNCS (LNAI), vol. 3626. Springer, Heidelberg (2005). https://doi.org/10.1007/978-3-540-31881-1
7. Priss, U.: Formal concept analysis in information science. Annu. Rev. Inf. Sci. Technol. (ASIST) **40**, 521–543 (2008)
8. Wolff, K.E.: A first course in formal concept analysis: how to understand line diagrams. Adv. Stat. Softw. **4**, 429–438 (1993)
9. EMAGE Gene Expression Database. http://www.emouseatlas.org/emage/. Accessed Dec 2017
10. Theiler, K.: The House Mouse: Atlas of Embryonic Development. Springer Science & Business Media, Heidelberg (2013)
11. Kaufman, M.H.: The Atlas of Mouse Development, vol. 428. Academic Press, London (1992)
12. Yevtushenko, S.A.: System of data analysis "concept explorer". In: Proceedings of 7th National Conference on Artificial Intelligence KII-2000, pp. 127–134 (2000). (in Russian)
13. Morgunova, E., Taipale, J.: Structural perspective of cooperative transcription factor binding. Curr. Opin. Struct. Biol. **47**, 1–8 (2017)
14. Boyer, L.A., Lee, T.I., Cole, M.F., Johnstone, S.E., Levine, S.S., Zucker, J.P., Guenther, M.G., Kumar, R.M., Murray, H.L., Jenner, R.G., et al.: Core transcriptional regulatory circuitry in human embryonic stem cells. Cell **122**(6), 947–956 (2005)
15. Wang, X., Guda, C.: Computational analysis of transcriptional circuitries in human embryonic stem cells reveals multiple and independent networks. BioMed. Res. Int. **2014** (2014)
16. Andrews, S., Hirsch, L.: A tool for creating and visualising formal concept trees. In: CEUR Workshop Proceedings: Proceedings of 5th Conceptual Structures Tools & Interoperability Workshop (CSTIW 2016), vol. 1637, pp. 1–9 (2016)
17. Dau, F., Andrews, S.: Combining business intelligence with semantic technologies: the CUBIST project. In: Hernandez, N., Jäschke, R., Croitoru, M. (eds.) ICCS 2014. LNCS (LNAI), vol. 8577, pp. 281–286. Springer, Cham (2014). https://doi.org/10.1007/978-3-319-08389-6_23
18. Andrews, S., McLeod, K.: Gene co-expression in mouse embryo tissues. Int. J. Intell. Inf. Technol. (IJIIT) **9**(4), 55–68 (2013)
19. Pensa, R.G., Boulicaut, J.-F.: Towards fault-tolerant formal concept analysis. In: Bandini, S., Manzoni, S. (eds.) AI*IA 2005. LNCS (LNAI), vol. 3673, pp. 212–223. Springer, Heidelberg (2005). https://doi.org/10.1007/11558590_22
20. Melo, C., Le-Grand, B., Aufaure, M.A., Bezerianos, A.: Extracting and visualising tree-like structures from concept lattices. In: Proceedings of 15th International Conference on Information Visualisation, pp. 261–266. IEEE (2011)
21. Taylor, A., McLeod, K., Burger, A.: Semantic visualisation of gene expression information. In: Andrews, S., Dau, F. (eds.) Proceedings of 3rd CUBIST Workshop. CEUR Workshop Proceedings, vol. 1040. CEUR (2013)

22. Dau, F.: Towards scalingless generation of formal contexts from an ontology in a triple store. Int. J. Concept. Struct. Smart Appl. (IJCSSA) 1(1), 18–38 (2013)
23. Kaytoue-Uberall, M., Duplessis, S., Napoli, A.: Using formal concept analysis for the extraction of groups of co-expressed genes. In: Le Thi, H.A., Bouvry, P., Pham Dinh, T. (eds.) MCO 2008. CCIS, vol. 14, pp. 439–449. Springer, Heidelberg (2008). https://doi.org/10.1007/978-3-540-87477-5_47
24. Blachona, S., Pensab, R.G., Bessonb, J., Robardetb, C., Boulicautb, J.F., Gandrillona, O.: Clustering formal concepts to discover biologically relevant knowledge from gene expression data. Silico Biol. 7, 476–483 (2007)
25. Kaytoue, M., Duplessis, S., Kuznetsov, S.O., Napoli, A.: Two FCA-based methods for mining gene expression data. In: Ferré, S., Rudolph, S. (eds.) ICFCA 2009. LNCS (LNAI), vol. 5548, pp. 251–266. Springer, Heidelberg (2009). https://doi.org/10.1007/978-3-642-01815-2_19
26. Kaytoue, M., Kuznetsov, S.O., Napoli, A., Duplessis, S.: Mining gene expression data with pattern structures in formal concept analysis. Inf. Sci. 181(10), 1989–2001 (2011)
27. Tarca, A.L., Romero, R., Draghici, S.: Analysis of microarray experiments of gene expression profiling. Am. J. Obstet. Gynecol. 195(2), 373–388 (2006)
28. Van't Veer, L.J., Dai, H., Van De Vijver, M.J., He, Y.D., Hart, A.A., Mao, M., Peterse, H.L., Van Der Kooy, K., Marton, M.J., Witteveen, A.T., et al.: Gene expression profiling predicts clinical outcome of breast cancer. Nature 415(6871), 530–536 (2002)

Exploring Heterogeneous Sequential Data on River Networks with Relational Concept Analysis

Cristina Nica, Agnès Braud, and Florence Le Ber[✉]

ICube, Université de Strasbourg, CNRS, ENGEES, Strasbourg, France
nica.cristina87@gmail.com, agnes.braud@unistra.fr,
florence.leber@engees.unistra.fr
http://icube-sdc.unistra.fr

Abstract. Nowadays, many heterogeneous relational data are stored in databases to be further explored for discovering meaningful patterns. Such databases exist in various domains and we focus here on river monitoring. In this paper, a limited number of river sites that make up a river network (seen as a directed graph) is given. Periodically, for each river site three types of data are collected. Our aim is to reveal user-friendly results for visualising the intrinsic structure of these data. To that end, we present an approach for exploring heterogeneous sequential data using Relational Concept Analysis. The main objective is to enhance the evaluation step by extracting heterogeneous closed partially-ordered patterns organised into a hierarchy. The experiments and qualitative interpretations show that our method outputs instructive results for the hydro-ecological domain.

1 Introduction

In Europe, according to the recomandations of Water Framework Directive [4], a special attention should be given to preserving or restoring the good state of waterbodies. Monitoring and assessing the effect of pollution sources and the one of restoration processes must be done in order to improve domain knowledge and to define guidelines for stakeholders.

During an interdisciplinary research project, namely REX[1], many and various hydro-ecological data have been collected periodically between 2002 and 2014 from a river network (seen as a directed graph). These data are about past restoration projects, temporal evolution of aquatic ecosystems and land use pressures. The REX data have been studied with statistical methods, but relational information could not be taken into account (e.g. effect of upper restoration).

Therefore, in this paper we deal with heterogeneous sequential data and we try to make sense of them by means of hierarchies of heterogeneous closed partially-ordered patterns (cpo-patterns, [2]) that exhibit the natural structure

---

[1] http://obs-rhin.engees.eu.

© Springer International Publishing AG, part of Springer Nature 2018
P. Chapman et al. (Eds.): ICCS 2018, LNAI 10872, pp. 152–166, 2018.
https://doi.org/10.1007/978-3-319-91379-7_12

of these data. Indeed, a cpo-pattern is compact, contains the same information as the set of sequential patterns it synthesises and is user-friendly thanks to its representation as a directed acyclic graph. Moreover, a hierarchy provides a convenient way for navigating to interesting heterogeneous cpo-patterns.

To that end, we extend our self-contained approach RCA-SEQ – introduced in [7] and based on Relational Concept Analysis (RCA, [12]) – for exploring classical sequential data to exploring heterogeneous sequential data. We propose to manipulate the data as a directed graph that has heterogeneous itemsets (i.e. a set of itemsets of different domains) as vertices and binary spatial relations as edges. Accordingly, we show that RCA-SEQ is robust and can be appropriate for exploring graphs and networks, as well.

The paper is structured as follows. Section 2 gives the theoretical background of our work. Section 3 describes the analysed heterogeneous hydro-ecological data. Section 4 introduces a data model used to encode the data into the RCA input. Section 5 presents the RCA-based exploration step. Section 6 explains our proposal for directly extracting hierarchies of heterogeneous cpo-patterns from the RCA output. In Sect. 7 experimental results are discussed. Section 8 presents related work. Finally, we conclude the paper in Sect. 9.

2 Preliminaries

2.1 Heterogeneous CPO-Patterns

Let $\mathcal{I} = \{I_1, I_2, \ldots, I_m\}$ be a set of *items*. An *itemset* $IS = (I_{j_1} \ldots I_{j_k})$, where $I_{j_i} \in \mathcal{I}$, is an unordered subset of $\mathcal{I}$. An itemset IS with k items is referred to as k-*itemset*. Let $\mathcal{IS}$ be the set of all itemsets built from $\mathcal{I}$. A *sequence* S is a non-empty ordered list of itemsets, $S = \langle IS_1 IS_2 \ldots IS_p \rangle$ where $IS_j \in \mathcal{IS}$. The sequence S is a *subsequence* of another sequence $S' = \langle IS'_1 IS'_2 \ldots IS'_q \rangle$, denoted as $S \preceq_s S'$, if $p \le q$ and if there are integers $j_1 < j_2 < \ldots < j_k < \ldots < j_p$ such that $IS_1 \subseteq IS'_{j_1}, IS_2 \subseteq IS'_{j_2}, \ldots, IS_p \subseteq IS'_{j_p}$.

Sequential patterns have been defined by [1] as frequent subsequences discovered in a sequence database. A sequential pattern is associated with a support θ, i.e. the number of sequences containing the pattern.

Suppose now that there is a *partial order* (i.e. a reflexive, antisymmetric and transitive binary relation) on the items, denoted by $(\mathcal{I}, \le)$. We say that $(\mathcal{I}, \le)$ is a *poset*. A *multilevel itemset* $IS_{ml} = (I_{j_1} \ldots I_{j_k})$, where $I_{j_i} \in \mathcal{I}$ and $\nexists I_{j_i}, I_{j_{i'}} \in IS_{ml}$ such that $I_{j_i} \le I_{j_{i'}}$, is a non-empty and unordered set of items that can be at different levels of granularity (i.e. items from different levels of poset $(\mathcal{I}, \le)$). We denote by $\mathcal{IS}_{ml}$ the set of all multilevel itemsets built from $(\mathcal{I}, \le)$. The partial order on the set of all multilevel itemsets $(\mathcal{IS}_{ml}, \subseteq_{ml})$ is defined as follows: $IS_{ml} \subseteq IS'_{ml}$ if $\forall I_j \in IS_{ml}, \exists I_{j'} \in IS'_{ml}, I_{j'} \le I_j$ and $\forall I_l \ne I_j, \exists I_{l'} \ne I_{j'}$ such that $I_{l'} \le I_l$.

To illustrate this, let us consider $\mathcal{I}_1 = \{a, b, c, d, e, Consonants, Vowels, Letters\}$ a set of items and $(\mathcal{I}_1, \le)$ a partial order depicted in Fig. 1, where an edge represents the binary relation *is-a*, denoted by $\le$.

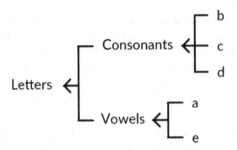

Fig. 1. An example of a partial order on $\mathcal{I}_1 = \{a, b, c, d, e,\ Consonants,\ Vowels,\ Letters\}$

For example, $a \leq Vowels$ designates that letter "a" is a vowel. Let be two itemsets $(a\ b\ c)$ and $(a\ Consonants)$, then $(a\ Consonants) \subseteq_{ml} (a\ b\ c)$ since $a \leq a$ and $b \leq Consonants$ (or $c \leq Consonants$).

Let $\mathcal{H} = \{\mathcal{I}_1, \mathcal{I}_2, \ldots, \mathcal{I}_n\}$ be a set of distinct sets of items, where $\mathcal{I}_j$ with $j \in \{1, \ldots, n\}$ represents a domain. We note that $\mathcal{I}_j$ can be a poset or an unordered set. Let $\mathcal{IS}_j$ be the set of all itemsets built from $\mathcal{I}_j \in \mathcal{H}$. A *heterogeneous itemset* $IS_{\mathcal{H}} = \{IS_1, IS_2, \ldots, IS_n\}$, where $IS_j \in \mathcal{IS}_j$, is a non-empty and unordered set of itemsets built from distinct sets of $\mathcal{H}$. In addition, a *multilevel heterogeneous itemset* (hereinafter referred to as heterogeneous itemsets) is a set of itemsets that has at least one multilevel itemset.

Let $\mathcal{IS}_{\mathcal{H}}$ be the set of all heterogeneous itemsets built from $\mathcal{H}$. The partial order $(\mathcal{IS}_{\mathcal{H}}, \subseteq_{\mathcal{H}})$ is defined as follows: $IS_{\mathcal{H}} \subseteq_{\mathcal{H}} IS'_{\mathcal{H}}$ if $\forall IS_k \in IS_{\mathcal{H}}, \exists IS'_k \in IS'_{\mathcal{H}}$ such that $IS_k \subseteq IS'_k$, where $IS_k, IS'_k \in \mathcal{IS}_k$, $k \in \{1, \ldots, n\}$. The order on heterogeneous itemsets is defined accordingly relying on $\subseteq_{ml}$.

To illustrate this, let us consider $\mathcal{H} = \{\mathcal{I}_1, \mathcal{I}_2\}$, where $\mathcal{I}_1$ is partially ordered as shown in Fig. 1 and $\mathcal{I}_2 = \{\square, \Diamond, \triangle\}$ is an unordered set of shapes. Furthermore, let be two multilevel heterogeneous itemsets $IS_{\mathcal{H}_1} = \{(Vowels\ c), (\Diamond)\}$ and $IS_{\mathcal{H}_2} = \{(a\ c), (\square\ \Diamond)\}$, then $IS_{\mathcal{H}_1} \subseteq_{\mathcal{H}} IS_{\mathcal{H}_2}$ since $(Vowels\ c) \subseteq_{ml} (a\ c)$ (that is $a \leq Vowels$ and $c \leq c$) and $(\Diamond) \subseteq (\square\ \Diamond)$.

A *heterogeneous sequence* $S_{\mathcal{H}} = \langle IS_{\mathcal{H}_1} IS_{\mathcal{H}_2} \ldots IS_{\mathcal{H}_r} \rangle$, where $IS_{\mathcal{H}_i} \in \mathcal{IS}_{\mathcal{H}}$ with $i \in \{1, \ldots, r\}$, is a non-empty ordered list of heterogeneous itemsets. In addition, a heterogeneous sequence that has at least one multilevel heterogeneous itemset represents a *multilevel heterogeneous sequence* (hereinafter referred to as heterogeneous sequence). A heterogeneous sequence $S_{\mathcal{H}}$ is a subsequence of another heterogeneous sequence $S'_{\mathcal{H}} = \langle IS'_{\mathcal{H}_1} IS'_{\mathcal{H}_2} \ldots IS'_{\mathcal{H}_q} \rangle$, denoted by $S_{\mathcal{H}} \preceq_{s_{\mathcal{H}}} S'_{\mathcal{H}}$, if $r \leq q$ and if there are integers $j_1 < j_2 < \ldots < j_k < \ldots < j_r$ such that $IS_{\mathcal{H}_1} \subseteq_{\mathcal{H}} IS'_{\mathcal{H}_{j_1}}, IS_{\mathcal{H}_2} \subseteq_{\mathcal{H}} IS'_{\mathcal{H}_{j_2}}, \ldots, IS_{\mathcal{H}_r} \subseteq_{\mathcal{H}} IS'_{\mathcal{H}_{j_r}}$. A frequent heterogeneous subsequence is called a *heterogeneous sequential pattern*.

To illustrate this, let be two heterogeneous sequences on the aforementioned $\mathcal{H} = \{\mathcal{I}_1, \mathcal{I}_2\}$: $S1_{\mathcal{H}} = \langle \{(a\ Consonants), (\square\ \Diamond)\}\ \{(Letters), \emptyset\} \rangle$ and $S2_{\mathcal{H}} = \langle \{(a\ d), (\square\ \triangle\ \Diamond)\}\ \{(a\ c), (\square)\} \rangle$. Then $S1_{\mathcal{H}} \preceq_{s_{\mathcal{H}}} S2_{\mathcal{H}}$ since

- $\{(a\ Consonants), (\square\ \Diamond)\} \subseteq_{\mathcal{H}} \{(a\ d), (\square\ \triangle\ \Diamond)\}$, i.e. $a \leq a$, $d \leq Consonants$, $(\square\ \Diamond) \subseteq (\square\ \triangle\ \Diamond)$,

– $\{(Letters), \emptyset\} \subseteq_{\mathcal{H}} \{(a\ c), (\square)\}$, i.e. $a \leq Letters$ (or $c \leq Letters$), $\emptyset \subseteq (\square)$.

Partially ordered patterns, *po-patterns*, have been introduced by [2], to synthesise sets of sequential patterns. Formally, a *po-pattern* is a directed acyclic graph $\mathcal{G} = (\mathcal{V}, \mathcal{E}, l)$. $\mathcal{V}$ is the set of vertices, $\mathcal{E}$ is the set of directed edges such that $\mathcal{E} \subseteq \mathcal{V} \times \mathcal{V}$, and l is the labelling function mapping each vertex to an itemset. With such a structure, we can determine a strict partial order on vertices u and v such that $u \neq v : u < v$ if there is a directed path from *tail* vertex u to *head* vertex v. However, if there is no directed path from u to v, these elements are not comparable. Each path of the graph represents a sequential pattern, and the set of paths in $\mathcal{G}$ is denoted by $\mathcal{P}_\mathcal{G}$. A po-pattern is associated to the set of sequences $\mathcal{S}_\mathcal{G}$ that contain all paths of $\mathcal{P}_\mathcal{G}$. The support of a po-pattern is defined as $Support(\mathcal{G}) = |\mathcal{S}_\mathcal{G}| = |\{S \in \mathcal{D}_S | \forall M \in \mathcal{P}_\mathcal{G}, M \preceq_s S\}|$. Furthermore, let $\mathcal{G}$ and $\mathcal{G}'$ be two po-patterns with $\mathcal{P}_\mathcal{G}$ and $\mathcal{P}_{\mathcal{G}'}$ their sets of paths. $\mathcal{G}'$ is a sub po-pattern of $\mathcal{G}$, denoted by $\mathcal{G}' \preceq_g \mathcal{G}$, if $\forall M' \in \mathcal{P}_{\mathcal{G}'}, \exists M \in \mathcal{P}_\mathcal{G}$ such that $M' \preceq_s M$. A po-pattern $\mathcal{G}$ is *closed*, referred to as *cpo-pattern*, if there exists no po-pattern $\mathcal{G}'$ such that $\mathcal{G} \prec_g \mathcal{G}'$ with $\mathcal{S}_\mathcal{G} = \mathcal{S}_{\mathcal{G}'}$. A cpo-pattern whose paths are heterogeneous sequential patterns is called *heterogeneous cpo-pattern*.

2.2 RCA

RCA extends the purpose of Formal Concept Analysis (FCA, [5]) to relational data. RCA applies iteratively FCA on a Relational Context Family (RCF). An RCF is a pair $(\mathcal{K}, \mathcal{R})$, where $\mathcal{K}$ is a set of object-attribute contexts and $\mathcal{R}$ is a set of object-object contexts. $\mathcal{K}$ contains n object-attribute contexts $K_i = (G_i, M_i, I_i), i \in \{1, \ldots, n\}$. $\mathcal{R}$ contains m object-object contexts $R_j = (G_k, G_l, r_j), j \in \{1, \ldots, m\}$, where $r_j \subseteq G_k \times G_l$ is a binary relation with $k, l \in \{1, \ldots, n\}$, $G_k = dom(r_j)$ the domain of the relation, and $G_l = ran(r_j)$ the range of the relation. G_k and G_l are the sets of objects of the object-attribute contexts K_k and K_l, respectively. RCA relies on a relational scaling mechanism that is used to transform a relation r_j into a set of *relational attributes* that extends the object-attribute context describing the set of objects $dom(r_j)$. A relational attribute $\exists r_j(C)$, where $\exists$ is the existential quantifier, and $C = (X, Y)$ is a concept whose extent contains objects from $ran(r_j)$, is owned by an object $g \in dom(r_j)$ if $r_j(g) \cap X \neq \emptyset$. Other quantifiers can be found in [12]. RCA process consists in applying FCA first on each object-attribute context of an RCF, and then iteratively on each object-attribute context extended by the relational attributes created using the learnt concepts from the previous step. The RCA result is obtained when the families of lattices of two consecutive steps are isomorphic and the object-attribute contexts are unchanged.

3 Heterogeneous Hydro-Ecological Data

We focus on hydro-ecological data concerning Rhine river. These data have been collected during REX project. A number of 15 *river sites* (i.e. fixed points) in the Alsace plain were monitored between 2002–2014. These sites make up the *river network* illustrated in Fig. 2 that can be seen as a graph of river sites linked by a spatial relation *is downstream of.*

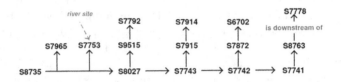

Fig. 2. River network

There are three monitored periods of time: 2002–2005 (I), 2006–2009 (II) and 2010–2014 (III). Periodically, for each river site a heterogeneous itemset {*physico-chemical* (PHC) *parameters, biological* (BIO) *indicators, land use*} is gathered. *PHC parameters* (e.g. temperature, nitrite and dissolved oxygen) indicate the presence or absence of different types of pollutions (e.g. organic or nutrient) according to the qualitative values of parameters. *BIO indicators* (e.g. Standardised Global Biological Index (IBGN), Biological Index of Diatoms (IBD) and Fish Biotic Index (IPR)) determine the quality of water. The indicators and parameters have five qualitative values provided by SEQ-Eau[2] standard, namely *very good, good, medium, bad* and *very bad* represented respectively by the colours *blue, green, yellow, orange* and *red*. All types of *land use* (e.g. forests and urban areas) effect positively or negatively the water quality. The land use around each monitored river site is assessed within two increasing buffers, precisely 100 m and 500 m. A type of land use, e.g. buildings, has a qualitative value according to a percentage of area j covered by it as follows: *low* if $j \in [0\%, 25\%]$, *medium* if $j \in (25\%, 52\%]$ and *high* if $j \in (52\%, 100\%]$. These domains are described by means of taxonomies as shown in Fig. 3. Let us note that the collected data concern only the atomic values from these taxonomies (e.g. urban areas).

In addition, a river site is included in a *river segment* that can be restored at one or more locations during the whole monitored period of time. There are two types of restoration: *global* and *wetland*. According to the number of restorations i undertaken during 2002–2014, there are three *levels of the type of restoration* as follows: $L1$ if $i \in (0, 2]$, $L2$ if $i \in (2, 5]$ and $L3$ if $i \in (5, \infty)$.

For instance, by analysing Fig. 4, the heterogeneous itemset {(NITRITE_{red}), (IBGN_{green}), (FORESTS_{low_500m} INDUSTRIAL AREAS_{high_500m})} is associated with river site *S7742* in period 2010–2014; the itemset (Wetland_{L1} Global_{L2}) is associated with river segment *20165* in the whole monitored period.

---

[2] http://rhin-meuse.eaufrance.fr/IMG/pdf/grilles-seq-eau-v2.pdf.

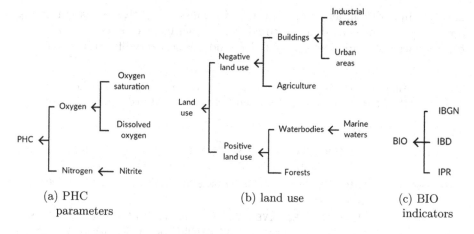

(a) PHC
   parameters

(b) land use

(c) BIO
   indicators

Fig. 3. Excerpts from the taxonomies over the three analysed domains

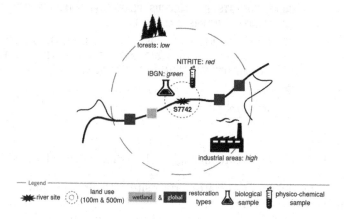

Fig. 4. River segment *20165*; period III: river site *S7742*

4   Data Modelling

Our purpose is to highlight how the ecological state of aquatic ecosystem and
the land use in upstream river sites impact the aquatic ecosystem in downstream
river sites and, thus determine the necessity of the restorations of river segments.
To that end, by exploiting the relational nature of the various collected hydro-
ecological data, we propose the data model shown in Fig. 5. This is used to
encode the analysed data into the RCA input. The six rectangles represent the
six sets of objects we manipulate: river segments, river sites, restoration types,
BIO indicators, PHC parameters and land use. These sets are given in Table 1.
The set of river sites contains all ordered pairs given by the Cartesian product
of the river sites shown in Fig. 2 and the three monitored periods of time. BIO
indicators, PHC parameters and land use correspond to the taxonomies depicted

in Fig. 3. In addition, let us mention that we consider, firstly, river segments as *target objects* since these are restored; and secondly, river sites as *non-target objects* since these are assessed to understand the necessity of restorations.

Table 1. Analysed sets of objects; the set of river sites is given by the Cartesian product of the river sites given in Fig. 2 and the three monitored periods of time

Set	Objects
River sites	{S7778, S8763, S7741, S6702, S7872, S7742, S7914, S7915, S7743, S7792, S9515, S8027, S7753, S7965, S8735} × {I, II, III}
River segments	3163, 4548, 5601, 6850, 8614, 8674, 18725, 19754, 19949, 20165, 20346, 26763
Land use	LAND USE, NEGATIVE LAND USE, POSITIVE LAND USE, BUILDINGS, AGRICULTURE, URBAN AREAS, INDUSTRIAL AREAS, LANDFILL & MINE SITES, ARABLE LANDS, PERMANENT CROPS, FORESTS & NATURAL AREAS, FORESTS, HERBACEOUS PLANTS, WETLANDS, WATERBODIES, CONTINENTAL WATERS, MARINE WATERS
PHC parameters	PHC, NITROGEN, NITRITE, AMMONIUM, PHOS, TOTAL PHOSPHORUS, NITRATE, OXYGEN, OXYGEN SATURATION, DISSOLVED OXYGEN, BIOLOGICAL OXYGEN DEMAND, TEMPERATURE
BIO indicators	IBGN, IBD, IPR
Restoration types	Wetland, Global

The links between objects are highlighted by using binary relations as follows:

– spatial relation *includes* associates a river site with a river segment if the river site is in the river segment;
– spatial relation *is downstream of* is used to encode into the RCA input the river network shown in Fig. 2;
– qualitative relation *has restoration L1/L2/L3* associates a river segment with the type of undertaken restoration;
– qualitative relation *has indicator blue/green/yellow/orange/red* associates a river site with a measured BIO indicator;
– qualitative relation *has parameter blue/green/yellow/orange/red* associates a river site with a measured PHC parameter;
– spatial-qualitative relation *is surrounded by low_100m/low_500m/medium_100m/medium_500m/high_100m/high_500m* associates a river site with a type of land use.

5   Exploration of Heterogeneous Data by Using RCA

In this section we briefly recall and slightly adapt the RCA-exploration step of sequential data presented in our previous paper [7].

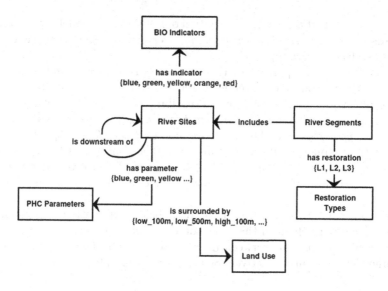

Fig. 5. Modelling heterogeneous hydro-ecological data

Firstly, the RCA input (RCF) – an excerpt is depicted in Table 2 – is built by relying on the data model shown in Fig. 5 and on the sets of objects given in Table 1. Basically, this RCF encodes all hydro-ecological data collected during the whole monitored period 2002–2014. There is an object-attribute context for each rectangle out of the data model, precisely KSegments (river segments), KSites (river sites), KRT (restoration types), KBIO (BIO indicators), KPHC (PHC parameters) and KLU (land use). KSites has no column since river sites are described only by using the *has indicator, has parameter* and *is surrounded by* relations. Similarly, KSegments has no column since river segments are described by using the *has restoration* relations. As shown in Table 2, a nominal scaling is used to build KRT in order to obtain a partial order over the unordered set of restoration types. In contrast, an ordinal scaling is used to build KBIO, KPHC and KLU in order to encode the taxonomies given in Fig. 3. In addition, there are 21

Table 2. Excerpt of the RCA input composed of object-attribute contexts (KSites, KRT and KPHC) and object-object contexts (RSite-ds-Site and RSite-red-BIO).

KSites
(S7743,I)
(S7743,II)
(S7743,III)
(S8735,I))
(S8735,II)

KRT	Wetland	Global
Wetland	×	
Global		×

KPHC	PHC	NITROGEN	NITRITE	PHOS
PHC	×			
NITROGEN	×	×		
NITRITE	×	×	×	
PHOS	×			×

RSite-ds-Site	(S7872,I)	(S7872,II)	(S9515,I)
(S7743,I)	×		
(S7743,II)		×	
(S7743,III)			
(S8735,I))	×		×
(S8735,II)		×	

RSite-red-PHC	PHC	NITROGEN	NITRITE
(S7743,I)			
(S7743,II)			×
(S7743,III)			
(S8735,I))			
(S8735,II)			×

object-object contexts, one for each relation out of the data model, e.g. in Table 2 `RSite-ds-Site` (river site *is downstream of* river site) and `RSite-red-BIO` (river site *has red* BIO *indicator*).

Secondly, RCA is applied[3] to the RCF shown in Table 2 and a family of concept lattices is obtained after four iterations. The RCA output comprises six concept lattices, one for each object-attribute context, as follows: *target lattice* $\mathcal{L}_{\text{KSegments}}$ (river segments), *non-target lattice* $\mathcal{L}_{\text{KSites}}$ (river sites), *lattice of restoration types* $\mathcal{L}_{\text{KRT}}$ and the *taxonomy lattices* $\mathcal{L}_{\text{KBIO}}$, $\mathcal{L}_{\text{KPHC}}$, $\mathcal{L}_{\text{KLU}}$ that correspond to the taxonomies illustrated in Fig. 3. The concepts of the latter three lattices are used to describe river sites by means of the revealed qualitative relational attributes. Similarly, the concepts of $\mathcal{L}_{\text{KRT}}$ are used to describe river segments.

It is worthwhile to mention that the RCA-based exploration step employs a relational scaling mechanism that relies on quantifier $\exists$ because the objective is to capture all the relations between the analysed objects. The target lattice and non-target one contain respectively 860 and 4554 concepts.

6   Extraction of Heterogeneous CPO-Patterns Organised into a Hierarchy

To extract a hierarchy of heterogeneous cpo-patterns from the RCA output (obtained as explained in Sect. 5), we apply and slightly modify the RCA-SEQ approach presented in [7]. Let us note that the hierarchy is directly obtained since each concept of the target lattice is associated with a heterogeneous cpo-pattern.

Briefly, starting with a concept from the target lattice, a heterogeneous cpo-pattern is extracted by navigating interrelated concept intents. For each navigated concept intent, a vertex (labelled with an itemset) is derived from all (spatial-)qualitative relational attributes (hereinafter referred to as qualitative relational attributes) whereas an edge is derived from a spatial relational attribute. A qualitative/spatial relational attribute highlights a qualitative/spatial relation.

In fact, in this paper a vertex derived from a concept intent of the non-target lattice (river sites) is actually a *heterogeneous vertex* labelled with a heterogeneous itemset. Basically, an itemset of the heterogeneous itemset is built for each set of qualitative relational attributes, which define the same qualitative relation, out of the concept intent. Therefore, for a concept intent we analyse the qualitative relational attributes, which are built using a qualitative relation r_q and concepts from a *taxonomy lattice* $\mathcal{L}_{K_{tax}} = (\mathcal{C}_{K_{tax}}, \preceq_{K_{tax}})$, to derive items as follows:

- from a qualitative relational attribute $\exists r_q(C_{tax})$, where $C_{tax} \in \mathcal{C}_{K_{tax}}$, is derived an item, denoted by "$item_q$", where $extent(C_{tax}) = \{item\}$ and q is the item quality according to r_q;

---

[3] Using http://dolques.free.fr/rcaexplore.

– if there is no qualitative relational attribute that highlights relation r_q and the information introduced by this relation is mandatory, then is derived an item, denoted by "$item_?$" where $extent(\top(\mathcal{L}_{K_{tax}})) = \{item\}$, that constitutes the 1-itemset obtained for this type of information; conversely, if the information introduced by this relation is not mandatory, then no item is derived and, thus $\emptyset$ is obtained for this type of information.

Let us mention that a vertex derived from a concept intent of the target lattice (river segments) is labelled with a multilevel itemset. As described in [7], this itemset can contain: the abstract item ("$?_?$" – different types of restoration at distinct number of locations), qualitative abstract items (e.g. "$?_{L1}$" – different types of restoration at most 2 locations) and/or concrete items (e.g. "$Global_{L3}$" – global restorations at more than 5 locations).

7 Results and Evaluation

In this section, we present some interesting results obtained with the RCA-SEQ approach applied to the heterogeneous data collected between 2002–2014 from the river network shown in Fig. 2. The evaluation relies on the positive feedback given by a hydro-ecologist who is well acquainted with cpo-patterns.

By navigating the lattices starting from the target concepts of $\mathcal{L}_{KSegments}$ we obtain a hierarchy of 859 heterogeneous cpo-patterns (the bottom concept of $\mathcal{L}_{KSegments}$ is not considered since generally it is too specific and associated with no river segment). It is worthwhile to mention that a smaller hierarchy of cpo-patterns can be extracted by varying the quantifiers employed by the relational scaling mechanism. In addition, various measures [6] can be used to select relevant heterogeneous cpo-patterns.

Figure 6 depicts an excerpt from this hierarchy, precisely the organised ①, ②, ③, ④, ⑤, ⑥ and ⑦ heterogeneous cpo-patterns.

A cpo-pattern is associated with a set of river segments whose number (support) is shown in ■. The restoration types of these river segments are illustrated in □ e.g. $Global_{L1}$ meaning that the river segments have at most 2 locations with global restoration. A vertex (○) is associated with a set of river sites and it is labelled with PHC parameters and their qualitative values. A vertex can have additional information: land use (○) and BIO indicators (◇). In the following, we focus on the cpo-patterns ①, ④ and ⑥.

CPO-Pattern ① is associated with 11 (■ in Fig. 6) river segments that contain at most 2 locations with global restoration. In addition, itemset (PHC_{blue}) reveals locally (i.e. in the associated river segments) a very good PHC state of water.

CPO-Pattern ④ is associated with 5 river segments that contain at most 2 locations with global restoration. Itemset (IBD_{green}) (◇ in Fig. 6) reveals locally a good ecological state of water based on the analysis of diatom species. In addition, the PHC state of water is very good for temperature, biological oxygen demand and nitrogen that represent a part of the abiotic characteristics suitable for diatom species [11].

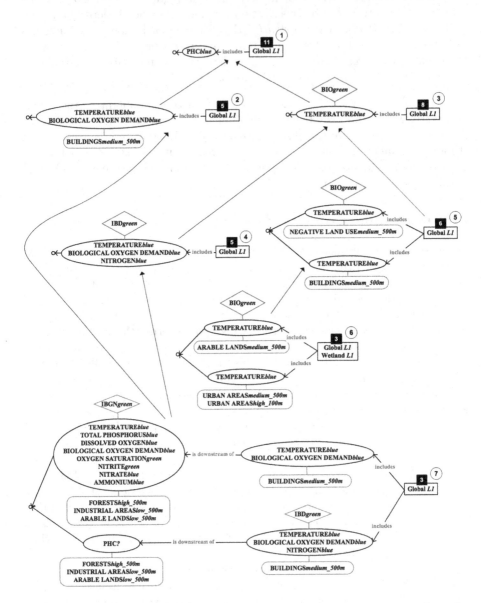

Fig. 6. Excerpt from the hierarchy of heterogeneous cpo-patterns discovered in the analysed hydro-ecological data. ①, ②, ③, ④, ⑤, ⑥ and ⑦ identify the cpo-patterns. ■ is the support (number of river segments) of a cpo-pattern; □ represents the types of river segment restoration; ◯ represents land use; ◊ represents BIO indicators; ◯ represents PHC parameters

CPO-Pattern ⑥, which is a more concrete specialisation of ⑤, is associated with 3 river segments that contain at most 2 locations with global and wetland restorations. Itemset (BIO$_\text{green}$) (◊ in Fig. 6) reveals locally a good ecological state of the aquatic ecosystem. Since BIO is an abstract item, we cannot specify the fauna and flora that underpin this regularity. In addition, itemset (TEMPERATURE$_\text{blue}$) reveals locally a very good PHC state of the water temperature. Furthermore, locally at 500 m buffer the land use pressures of arable lands and urban areas are *medium* whereas at 100 m the land use pressures of urban areas are *high*.

Figure 7 depicts a more complex heterogeneous cpo-pattern. This is associated with the river segments *8674* and *19949* that contain at most 2 locations with global restoration.

The vertices are derived from concepts of $\mathcal{L}_\text{KSites}$ whose extents are as follows: Ⓐ: {(*S7743*,III), (*S7915*, III)}, Ⓑ: {(*S7915*,II), (*S7743*,III)}, Ⓒ: {(*S7915*, I), (*S7743*,III)}, Ⓓ: {(*S7915*,II), (*S7743*,II)}, Ⓔ: {(*S7914*,III)} and Ⓕ: {(*S7914*,I), (*S7914*,II), (*S7914*,III)}. Locally, in the whole monitored period 2002–2014 the land use pressures of buildings are *medium* at 500 m buffer. In contrast, in the upstream rivers at 500 m buffer on the one hand the land use pressures of industrial areas and arable lands are *low*; on the other hand, a *high* percentage of the area is covered with forests that lead to a good ecological state of the aquatic ecosystem in the surroundings. Indeed, by analysing the Ⓔ vertex, itemset (IBGN$_\text{green}$) (◊, Fig. 7) reveals a good ecological state of the aquatic ecosystem in the period 2010–2014 based on the analysis of macro-invertebrates. Moreover, water temperature is very good; organic matter (dissolved oxygen, biological oxygen demand and oxygen saturation) is good and very good; nitrogenous parameters (nitrite and ammonium), which are related to organic matter, are as well good and very good; and nutrients (total phosphorous and nitrate) are very good.

By comparing the Ⓔ vertex with the Ⓐ, Ⓑ, Ⓒ and Ⓓ vertices, it is noted a degradation up to one level regarding the qualitative values of PHC parameters probably caused by the *medium* building pressures at 500 m buffer, e.g.:

- AMMONIUM$_\text{blue}$ and DISSOLVED OXYGEN$_\text{blue}$ (very good) from Ⓔ are measured when the surroundings are covered with a low percentage of industrial areas and arable lands (i.e. the land use pressures are low), while AMONIUM$_\text{green}$ and DISSOLVED OXYGEN$_\text{green}$ (good) from Ⓐ, Ⓑ, Ⓒ and Ⓓ are measured when the surroundings are covered with a medium percentage of buildings (i.e. the land use pressures are medium);
- TOTAL PHOSPHORUS$_\text{blue}$ (very good) from Ⓔ is measured when in the surroundings the land use pressures are low; TOTAL PHOSPHORUS$_\text{green}$ (good) from Ⓑ, Ⓒ and Ⓓ is measured when in the surroundings the land use pressures are medium.

Furthermore, the cpo-pattern shown in Fig. 7 reflects that BIO indicators seem to be more sensitive (up to two levels of their qualitative values) to land use pressures [14,15]. For instance, IBGN$_\text{green}$ in upstream rivers (Ⓔ) in contrast to BIO$_\text{yellow}$ and IBGN$_\text{orange}$ locally (Ⓒ and Ⓓ, respectively).

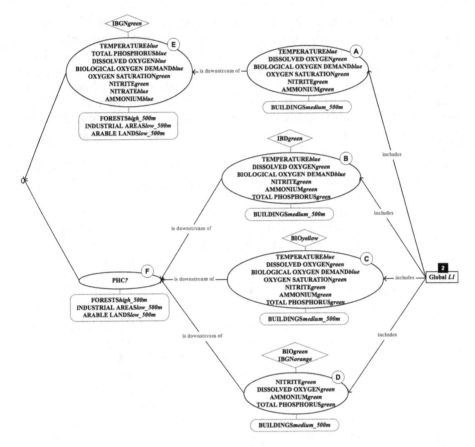

Fig. 7. A complex heterogeneous cpo-pattern extracted from the analysed hydro-ecological data. Ⓐ, Ⓑ, Ⓒ, Ⓓ, Ⓔ and Ⓕ identify the vertices; ■ is the support (number of river segments) of the cpo-pattern; □ represents the types of river segment restoration; ○ represents land use; ◇ represents BIO indicators; ○ represents PHC parameters

To sum up, by means of cpo-patterns, we can help hydro-ecologists to check well-known correspondences among the analysed ecological factors as well as to consider lesser-known facts.

8   Related Work

Classical sequential pattern mining approaches deal with sequences whose items are homogeneous and, therefore cannot be applied to heterogeneous sequences (i.e. sequences whose items are different in nature). To our knowledge, [9] proposed the first work for exploring multidimensional sequential data. A multidimensional sequence takes the form $(d_1, d_2, \ldots, d_m, S)$, where S is a sequence of itemsets and d_i represents the ith type of information associated with S. The authors proposed three methods for extracting multidimensional sequential patterns that rely on classical sequential pattern algorithms (e.g. PREFIXSPAN [8]). A key drawback of such multidimensional sequences is the additional information that is constant for all itemsets of sequence S.

In [10], a multidimensional sequence is defined as an ordered list of multidimensional items. A multidimensional item takes the form $(d_1, d_2, \ldots, d_n)$, where d_k is an item of the kth dimension. Furthermore, each considered dimension is represented at different levels of granularity by means of partial orders. Therefore, multilevel sequential patterns can be discovered, as explained in [13]. The authors proposed the M3SP algorithm that searches for multidimensional and multilevel sequential patterns in two steps. First, the most specific frequent multidimensional items, referred to as *maf-sequences*, are found. Second, the maf-sequences are used to remodel the original multidimensional sequences and then these sequences are mined by using algorithm SPADE [16].

Nevertheless, [3] highlighted a limitation of M3SP, i.e. the multidimensional items do not allow itemsets whose items are of kth dimension. The authors proposed the MMISP algorithm that considers complex and heterogeneous sequences, where a sequence contains *elementary sequences* (ES), i.e. itemsets whose items can be of two types: atomic and different in nature taken from user-defined taxonomies or subsets of unordered sets of items. MMISP does not discover directly sequential patterns in heterogeneous data since a preprocessing step is involved, i.e. the original sequences are encoded into classical sequences. In contrast, RCA-SEQ directly searches for cpo-patterns (rather than sequential patterns) in complex and heterogeneous data and, besides, reveals how these patterns relate to each other. Moreover, our approach generalises the ES proposed in [3] by considering its atomic items as 1-itemsets.

9   Conclusion

RCA-SEQ is an approach for exploring classical sequential data. In this paper, we have presented an extension of RCA-SEQ that highlights its generality, i.e. the capability to explore sequential data regardless of their complexity. Given heterogeneous sequential data on river networks, we have shown that hydro-ecologists can draw valuable insights by exploiting the "richness" (e.g. the additional information captured by concept extents and the revealed abstract items) of the RCA-SEQ output. In the future, we plan to (i) improve our extension in order to be applicable to large volumes of heterogeneous sequential data and (ii) explore with RCA-SEQ complex relational data such as social networks and knowledge graphs.

Acknowledgement. The REX data were provided by Prof. Jean-Nicolas Beisel and the interpretation of cpo-patterns was done with the help of Corinne Grac (ENGEES Strasbourg).

References

1. Agrawal, R., Srikant, R.: Mining sequential patterns. In: International Conference on Data Engineering, pp. 3–14 (1995)
2. Casas-Garriga, G.: Summarizing sequential data with closed partial orders. In: 2005 SIAM International Conference on Data Mining, pp. 380–391 (2005)
3. Egho, E., Jay, N., Raïssi, C., Ienco, D., Poncelet, P., Teisseire, M., Napoli, A.: A contribution to the discovery of multidimensional patterns in healthcare trajectories. J. Intell. Inf. Syst. **42**(2), 283–305 (2014)
4. European Union: Directive 2000/60/EC of the European parliament and of the council of 23 October 2000 establishing a framework for community action in the field of water policy. Off. J. OJ L **327**, 1–73 (2000)
5. Ganter, B., Wille, R.: Formal Concept Analysis: Mathematical Foundations. Springer, Heidelberg (1999). https://doi.org/10.1007/978-3-642-59830-2
6. Nica, C., Braud, A., Dolques, X., Huchard, M., Ber, F.L.: Exploring temporal data using relational concept analysis: an application to hydroecological data. In: Proceedings of the 13th International Conference on Concept Lattices and Their Applications, CLA 2016, pp. 299–311. CEUR-WS.org (2016)
7. Nica, C., Braud, A., Dolques, X., Huchard, M., Le Ber, F.: Extracting hierarchies of closed partially-ordered patterns using relational concept analysis. In: Haemmerlé, O., Stapleton, G., Faron Zucker, C. (eds.) ICCS 2016. LNCS (LNAI), vol. 9717, pp. 17–30. Springer, Cham (2016). https://doi.org/10.1007/978-3-319-40985-6_2
8. Pei, J., Han, J., Mortazavi-Asl, B., Pinto, H., Chen, Q., Dayal, U., Hsu, M.C.: PrefixSpan: mining sequential patterns efficiently by prefix-projected pattern growth. In: Proceedings of the 17th International Conference on Data Engineering, ICDE 2001, pp. 215–224. IEEE Computer Society (2001)
9. Pinto, H., Han, J., Pei, J., Wang, K., Chen, Q., Dayal, U.: Multi-dimensional sequential pattern mining. In: Proceedings of the 10th International Conference on Information and Knowledge Management, CIKM, pp. 81–88. ACM (2001)
10. Plantevit, M., Laurent, A., Laurent, D., Teisseire, M., Choong, Y.W.: Mining multidimensional and multilevel sequential patterns. ACM Trans. Knowl. Discov. Data **4**(1), 4:1–4:37 (2010)
11. Raibole, M., Singh, Y.: Impact of physico-chemical parameters on microbial diversity: seasonal study. Curr. World Environ. **6**(1), 71–76 (2011)
12. Rouane-Hacene, M., Huchard, M., Napoli, A., Valtchev, P.: Relational concept analysis: mining concept lattices from multi-relational data. Ann. Math. Artif. Intell. **67**(1), 81–108 (2013)
13. Srikant, R., Agrawal, R.: Mining sequential patterns: generalizations and performance improvements. In: Apers, P., Bouzeghoub, M., Gardarin, G. (eds.) EDBT 1996. LNCS, vol. 1057, pp. 1–17. Springer, Heidelberg (1996). https://doi.org/10.1007/BFb0014140
14. Villeneuve, B., Souchon, Y., Usseglio-Polatera, P., Ferréol, M., Valette, L.: Can we predict biological condition of stream ecosystems? A multi-stressors approach linking three biological indices to physico-chemistry, hydromorphology and land use. Ecol. Ind. **48**, 88–98 (2015)
15. Wasson, J., Villeneuve, B., Mengin, N., Pella, H., Chandesris, A.: Quelle limite de "bon état écologique" pour les invertébrés benthiques en rivières? Apport des modèles d'extrapolation spatiale reliant l'indice biologique global normalisé à l'occupation du sol. Ingénieries - EAT **1**(47), 3–15 (2006)
16. Zaki, M.J.: Spade: an efficient algorithm for mining frequent sequences. Mach. Learn. **42**(1), 31–60 (2001)

Conceptual Graphs Based Modeling
of Semi-structured Data

Viorica Varga[(✉)], Christian Săcărea, and Andrea Eva Molnar

Babes-Bolyai University, str. Kogalniceanu 1, 400081 Cluj Napoca, Romania
ivarga@cs.ubbcluj.ro, {csacarea,andrea.molnar}@math.ubbcluj.ro

Abstract. Due to the fast growing of data in the digital world, not only in volume but also in its variety (structured, un-structured or hybrid), traditional RDBMS are complemented with a rich set of systems, known as NoSQL. One of the main categories of NoSQL databases are document stores which are specifically designed to handle semi-structured data, for instance XML documents. In this paper, we present a modeling method for semi-structured data based on Conceptual Graphs and exemplify the method on an XML document. The expressive power of Conceptual Graphs makes them particularly suitable for conceptual modeling of semi-structured data.

1   Introduction and Related Work

NoSQL data stores vary not only in their data, but also in their query model. The most common categorization of these systems is by their data model, distinguishing key-value stores, document stores, column-family stores, and graph databases [1]. Document based NoSQL systems are based on the semi-structured data model, but most NoSQL data stores do not enforce any structural constraints on the data; they are usually referenced as schema-less data. On the other hand, for managing and retrieving data, its inherent structure proves to be significant. Not knowing the general structure of the data, makes tasks like application development or data analysis very difficult. Especially for modeling purposes, conceptual design of semi-structured data proves to be an important task. Several visualization methods have been developed over time in order to enhance understanding and to offer reasoning support for non-experts. For instance, Visual Query Systems (VQS) give a visual solution for non expert users which no longer have to understand query languages such as SQL or XQuery. A survey on visual query systems is given in [4] with the purpose of facilitating querying databases to non expert users. For instance, the modeling language of Conceptual Graphs (CG) can be used to model relational database design and querying [11]. In this research, we continue the ideas developed in [11] and describe how CG can be employed for conceptual modeling of semi-structured data.

Originally, CGs have been introduced by Sowa in [8] to model database interfaces and then further elaborated in [9]. Since then, CG have been used as a

© Springer International Publishing AG, part of Springer Nature 2018
P. Chapman et al. (Eds.): ICCS 2018, LNAI 10872, pp. 167–175, 2018.
https://doi.org/10.1007/978-3-319-91379-7_13

conceptual schema language, as well as a knowledge representation language with goal to provide a graphical representation for logic which is able to support human reasoning. Among the plethora of interesting applications, we would like to mention a consistent, graphical interface for database interaction. CG have been successfully mathematically formalized for representing database structures in [2,3].

A unified CG approach to represent database schema - including relations between tables, and to model queries in databases has been described in [11]. XQuery, the standard language to query XML data, is gaining increasing popularity among computer scientists. A representation method of XQuery using CG which proves to be a good visual tool for even for non-experts has been discussed in [5,6]. The structure of the XML data is given as a CG, helping the user to construct the query on the selected data. A graphical user interface BBQ (Blended Browsing and Querying) is proposed in [7] for browsing and querying XML data sources. BBQ is a query language for XML-based mediator systems (a simplification of XML-QL) which allows queries incorporating one or more of the sources. In BBQ XML elements and attributes are shown in a directory-like tree and the users specify possible conditions and relationships (as joins) among elements.

In this paper, we continue our previous research on modeling various databases using FCA and CG [10,11] and we propose a CG based modeling of semi-structured data. This conceptual model can be useful both for document-oriented database design and for XML data modeling. We exemplify the developed methods on a toy XML data set. This research come along with a software tool for XML data design using CG. This tool has been developed targeting a wide range of users, from researchers wanting to validate their work using the CG grounded conceptual model developed in this paper to non-experts, who would like to rely on the expressive power of CG as a visual interface, which gives the user the possibility to formulate a specific hierarchical view, without any knowledge of the detailed structure and the content of the database constructs.

2   Graphic Representation of Semi-structured Data Model by Conceptual Graphs

In the following, we propose a CG grounded representation of semi-structured data model. The method is similar to the representation with E/R diagrams and consists of the following three main steps:

Step 1: Identify all complex objects. Every object will be modeled as a CG **concept** and graphically represented as a rectangle.

Step 2: Define the relationships between the objects. The relationship between two or more **concepts** will be represented by means of a **relation** and graphically represented as an oval. Similarly to the approach in [8], a directed arc from the first concept node to the relation node and another arc from the relation to the second concept node will represent this relationship. If a relation has only one element, the arrowhead is omitted. If a relation has more than two elements, the arrowheads are replaced by integers $1, \ldots, n$. The **direction of arcs are from the root to leaf** levels. According to the type of the relationship between the objects, a relation can be:

- `hasOne`: models a one-to-one relationship between two objects.
- `hasMore`: illustrates a one-to-many relationship.
- `hasKey`: specifies the key object of a given object;
- `refers`: represents a reference between two or more object. It can be considered as a graphical representation of the foreign key constraint: the directed arc points from the foreign key to the key element.
- `isOptional`: designs the case when the minimum number of occurrences of an object is zero. If this relation is omitted, the default value of the number of occurrences (of the corresponding object) is one.

The structure of the semi-structured data gives a natural representation of the *one-to-many* relationships. The most common representation is that the child object is embedded into the parent object. The advantage of this modeling is that we do not have to perform a separate query in order to get the details of the embedded data, so the execution time of a query will decrease. The disadvantage is that we cannot access the embedded details as independent entities.

By contrast, the modeling of *many-to-many* relationship is particularly challenging in XML, but there exist a couple of possible solutions to model it: nested in one of the parents and one foreign key referencing to the other parent, or as in relational databases, at root level with two foreign keys. The most common representations use the first possibility. Figure 1 presents the way of designing the relationship between two related objects. Let O_1 and O_2 be two complex objects. We assume that O_{1n} is the key object (represented by the relation `hasKey`) of O_1. Object O_{21} (as foreign key) of O_2 references O_{1n}, represented by the relation `refers`.

Step 3: Identify all descendant objects of the complex objects, defined at *Step 1*. Remark that, in the case of Document Store Databases there is no possibility to define the type of the elementary attributes. Hence, this step will be presented in detail in the next paragraph, concerning to the XML Data Model.

3   CG Based Representation of XML Data Structure

Similar to the above concept, we present in the following how XML data structure can be represented using CGs (referenced as XML Schema CG in the following).

Step 1. Identify all complex elements. The **concept type** can represent: an *element* (E); an *attribute* (A); a *root element* (R); the *type of the data* (T); an *enumeration constraint* (N). At the same time, the **concept referent** can be:

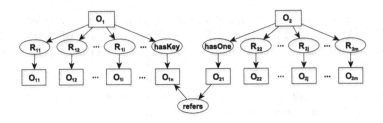

Fig. 1. CG describing a connection between the related objects O_1 and O_2.

- the *name of* the corresponding *node* (element or attribute);
- the *data type of a simple element* or *an attribute*;
- the *set of acceptable values* of the corresponding *element*, in the case of enumeration. In this particular case, the concept name will be composed by the predefined values of the enumeration, separated by ($|$).

Step 2. Define the relationship between the nodes. As we defined earlier, the relationship between two or more **concepts** will be illustrated by a CG **relation**. The relations presented in the previous section are adaptable also to the XML Schema documents. Hence, we analyze only the additional relations:

- `refers`: as in the general case, it represents a linkage between two or more complex elements. In the case of XML Schema, this relation can express the ID/IDREF couples with the help of the *hasKey* relation.
- `hasType`: defines the data type of a simple element or an attribute. If the *hasKey* relation is set, this relation can be omitted in the case of the corresponding element or attribute.
- `hasChoice` and `isPossibility`: specify the child elements of a choice compositor. In the XML Schema representation, the definition of a choice element allows only one of the elements contained in the declaration to be present within the containing element.
- `isEnum`: represents the enumeration constraint, which limits the content of an element to a set of acceptable values.

Step 3. Identify all child elements or attributes of the complex elements, defined at *Step 1*, give the data type for each one and specify also the potential restrictions of the elements or attributes.

4   An Example to Illustrate the Representation of XML Schema with Conceptual Graphs

In this section, we present the previously detailed methods on a graphical representation of XML Schema using CG through a detailed example. Let us consider the structure of data in the case of a university. The data is stored in XML documents, but we want a graphical representation which illustrates more profoundly the hierarchical structure of our data.[1]

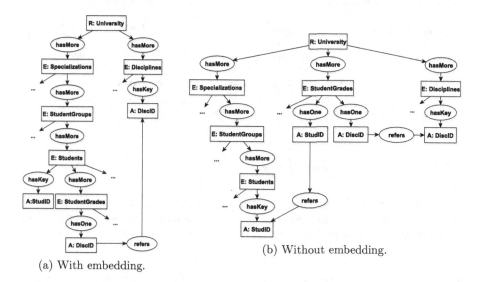

(a) With embedding.

(b) Without embedding.

Fig. 2. Two possible solutions for designing the many-to-many relationships between the **Students** and **Disciplines** elements.

Step 1. We define a concept, called **University**, which will represent the root element of the XML Schema. The **University** element contains some different child elements, e.g. **Specializations** and **Disciplines**, which are all complex elements. So, let us construct their descendants and let us define the relationships among them.

Step 2. Students of a specialization are organized in groups. Therefore, we have three complex elements in the graph as concepts according to the following hierarchy: **Specialization** becomes the root element, while **StudentGroups** and **Students** become child elements of **Specialization** and **StudentGroups**, respectively; this parent-child relationship will be represented by **hasMore** relations. Now, let us construct the elements **Disciplines** and **StudentGrades**. Notice that students have grades for more disciplines. Hence, there exists a many-to-many relationship between **Students** and **Disciplines**, which is represented by the **StudentGrades** element. Figures 2a and b show two main possible hierarchies, which can be formulated for this many-to-many relationship. Figure 2a designs the classic solution: the **StudentGrades** element is nested into

---

[1] Due to obvious space reasons, we have not included the complete XML schema (referred as *UniversityXSD* in the following) and the corresponding XML document, but they can be consulted at http://www.cs.ubbcluj.ro/~fca/semistructured-data/.

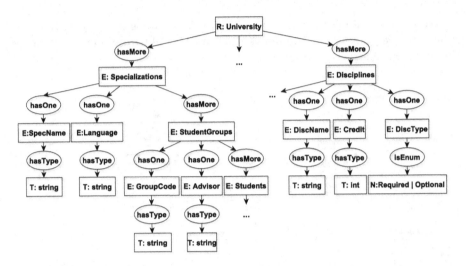

Fig. 3. The part of the University XML Schema CG emphasizing the hierarchy between the Specializations, StudentGroups and Students elements and showing the structure of the Specializations, StudentGroups and Disciplines elements.

the Students element. Hence, the relationship between these two elements is defined. The linkage between Disciplines and StudentGrades elements is represented by the dyadic relation refers, which links the DiscID (as "foreign key") from StudentGrades with the *key* element of Disciplines element, i.e. the key attribute DiscID. Remark that, in the *UniversityXSD* XML Schema the many-to-many relationship is represented in this standard way. Another possibility to illustrate this type of relationship would be embedding the StudentGrades element into the Disciplines element. This representation is identical to the above solution, so we omit its graphical representation.

Figure 2b shows the third possible method for designing the many-to-many relationship. The StudentGrades element is selected as the child of the root element. In this situation, the StudentGrades element links the Students element with Disciplines element through the reference elements StudID and DiscID, respectively. Remember that, in the case of ID/IDREF couples the use of hasType relation and the data type concept, respectively, could be omitted (see the attributes StudID and DiscID). Notice also that the name of the connecting nodes can be different in the parent and child elements.

Step 3. A detailed design for the complex elements is illustrated in Figs. 3 and 4. Let us analize the Students element (see Fig. 4). We have two possibilities to define the StudentName element, using the hasChoise relation: we can give the full name of the student as one string or we can give the parts of the name separately. According to the definition of the choice element, one of the possibilities must be selected. Element BirthPlace is selected as optional. Notice that, a hasMore relation can connect a complex element with a simple element, e.g.

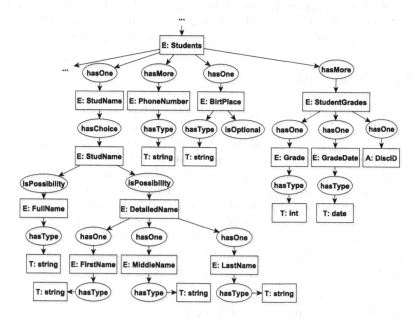

Fig. 4. The snippet of the **University** XML Schema CG illustrating the structure of the **Students** and **StudentGrades** elements.

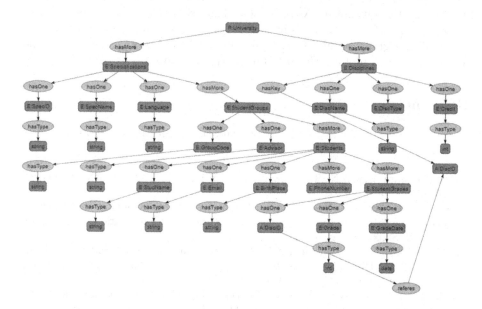

Fig. 5. Application screen-shot for UnivesityXSD design

the relation between `Students` and `PhoneNumber` elements. In the case of the `Disciplines` element it is defined an enumeration constraint by the `DiscType` element, see Fig. 3: we can choose between *Required* and *Optional*.

5 The Visual Interface

We have implemented a tool, named *XSD Builder*, which provides a user-friendly interface in form of CGs and gives the user the possibility to build the graphs of personalized XML Scheme. A snippet from the structure of the detailed XML Schema example, presented in the previous section (*UniversityXSD*) visualized by *XSD Builder* can be seen in Fig. 5. The user can select one concept or relation from the toolbox, by dragging it to the schema pane. Then, the user can give the name of the selected object, by writing it into the rectangle or the oval. It is also easy to specify the relationship between the concepts and relations: the user can draw a line between them. The tool gives the possibility to generate XML Scheme from the CG designed in it with a single click. Also, given an XSD file, the application will visualize the corresponding CG.

6 Conclusions and Future Work

The goal of CGs is to provide a graphical representation for logic which is able to support human reasoning. This article proposes an application which provides a graphical interface for Semi-Structured Data Design in form of CGs. We implemented a software for XML Data Design using CG. Our future goal is to extend the list of available tools in the case of XML schema representation (e.g. involving the group element). The presented representation using CG can be implemented as a tool for MongoDB data structure too.

References

1. Cattell, R.: Scalable SQL and NoSQL data stores. SIGMOD Rec. **39**(4), 12–27 (2011)
2. Dau, F.: The Logic System of Concept Graphs with Negation. And Its Relationship to Predicate Logic. LNCS, vol. 2892. Springer, Heidelberg (2003). https://doi.org/10.1007/b94030
3. Dau, F., Hereth, J.C.: Nested concept graphs: mathematical foundations and applications for databases. In: Using Conceptual Structures. Contributions to ICCS, pp. 125–139. Shaker Verlag, Aachen (2003)
4. Lloret-Gazo, J.: A survey on visual query systems in the web era. In: Hartmann, S., Ma, H. (eds.) DEXA 2016. LNCS, vol. 9828, pp. 343–351. Springer, Cham (2016). https://doi.org/10.1007/978-3-319-44406-2_28
5. Molnar, A., Varga, V., Sacarea, C.: Conceptual graph driven modeling and querying methods for RDMBS and XML databases. In: Proceedings of the 13th International Conference on Intelligent Computer Communication and Processing, ICCP 2017, pp. 55–62. IEEE (2017)

6. Molnar, A., Varga, V., Sacarea, C.: Conceptual graphs based modeling and querying of XML data. In: Proceedings of the 25th International Conference on Software, Telecommunications and Computer Networks, SoftCOM, pp. 23–28. IEEE (2017)
7. Munroe, K.D., Papakonstantiou, Y.: BBQ: a visual interface for integrated browsing and querying of XML. In: Arisawa, H., Catarci, T. (eds.) Advances in Visual Information Management. ITIFIP, vol. 40, pp. 277–296. Springer, Boston, MA (2000). https://doi.org/10.1007/978-0-387-35504-7_18
8. Sowa, J.F.: Conceptual graphs for a database interface. IBM J. Res. Dev. **20**(4), 336–357 (1976)
9. Sowa, J.F.: Conceptual Structures: Information Processing in Mind and Machine. Addison-Wesley, Boston (1984)
10. Varga, V., Janosi-Rancz, K.T., Kalman, B.: Conceptual design of document NoSQL database with formal concept analysis. Acta Polytech. Hung. **13**(2), 229–248 (2016)
11. Varga, V., Sacarea, C., Takacs, A.: Conceptual graphs based representation and querying of databases. In: Proceedings of the International Conference on Automation, Quality and Testing, Robotics (AQTR), vol. 03, pp. 1–6. IEEE (2010)

Using Conceptual Structures
in Enterprise Architecture to Develop
a New Way of Thinking and Working
for Organisations

Simon Polovina[1,2(✉)] and Mark von Rosing[2]

[1] Conceptual Structures Research Group, Communication and Computing Research Centre and Department of Computing, Sheffield Hallam University, Sheffield, UK
S.Polovina@shu.ac.uk
[2] Global University Alliance, Chateau Du Grand Perray, La Bruere Sur Loir, France
mvr@globaluniversityalliance.org

Abstract. Enterprise Architecture (EA) is a discipline that provides generic patterns that any organisation can reuse throughout its own business, informatics and technical components. However, EA's current way of thinking and working to achieve this aim is not standardised. EA thus continues to "reinvent the wheel" that causes mistakes or wastes resources on rediscovering what should already be known. We, therefore, represent the specific business, information and technology meta-models as patterns that can be fully reintegrated in one repeatable meta-model for the whole organisation. The outcome is a new agile way of thinking and working, highlighted by how EA works better in enterprise layers, sub-layers and levels of abstraction. To test the meta-models, two forms of Conceptual Structures known as Conceptual Graphs (CGs) and Formal Concept Analysis (FCA) are brought together through the *CGtoFCA* algorithm. The algorithm identifies how the layered meta-models can share meaning and truth and without having to recombine them into one large, unwieldy meta-model as the repeatable structure.

1 Introduction

Organisations can draw upon leading and best practices to gain insight into how best to fulfil their value and purpose. This insight ranges from understanding the external forces (e.g. the marketplace or non-profit environment) and internal forces (e.g. career ambitions of their employees) from which the organisations derive their strategy. The insight ranges to the operational behaviour (e.g. business processes), computer-based applications and data needed to implement that strategy most effectively.

Enterprise Architecture (EA) is a discipline that provides generic patterns that any organisation can reuse throughout its own business, informatics and data models in fulfilment of that organisation's overall purpose. The organisation

© Springer International Publishing AG, part of Springer Nature 2018
P. Chapman et al. (Eds.): ICCS 2018, LNAI 10872, pp. 176–190, 2018.
https://doi.org/10.1007/978-3-319-91379-7_14

thus avoids "reinventing the wheel" which causes it to make mistakes or waste resources on rediscovering what is already known.

To reduce misinterpretation, the patterns are intended as formal models of the models—i.e. meta-models—that each business can specialise according to their specific needs. Computer science and informatics contributes to the express-ibility in these meta-models through its advances in ontology and semantics; together they capture the objects and relations that describe the interplay and effects of business in a formal, computable model [2,3].

There are however multiple EA frameworks and methods, each with their own meta-models and associated approaches revealing a lack·of mutual under-standing what the meta-models should consist of and how they ought to be used. The content within and interconnections between the meta-models for the 'architectural domains'—i.e. business, information and technology—that make up the organisation are also interpreted differently according to the EA frame-work. The inconsistencies in the meta-models, and how to think and work with them undermine our conceptual understanding of organisations with potentially damaging effect. Consequently, organisations still end up reinventing the wheel.

The classical way of thinking and working in EA's architectural domains with a linear waterfall approach is counterproductive to the aims of EA. We evidence that representing the architectural domains as 'layers' enables us to think and work simultaneously within and across these multiple domains. We test this approach through Conceptual Structures, namely Conceptual Graphs and Formal Concept Analysis. As well as offering an agile way of thinking and working with EA, organisations can thus better draw upon the suggested best and leading practices to gain insight into how best to fulfil their value and purpose.

2   Understanding Architectural Layers in Organisations

Independent of their size or industry, organisations share a common underlying structure that consists of the following enterprise layers identified from previous work [8]:

- Business: Such as the purpose and goal, competencies, processes, and services aspects;
- Information: Such as the application systems, as well as the data components;
- Technology: Such as the platform and infrastructure components.

These layers represent the three perspectives by which organisations are viewed. They are called layers because the business layer sits on top of the information layer that in turn sits on top of the technology layer. It epitomises that organisations are driven by their business needs that are enabled by their information systems (applications and data), which require the underlying tech-nology to run these systems.

The organisation thus has to align its way of thinking with its way of working within and across all these perspectives. The Global University Alliance (GUA, www.globaluniversityalliance.org) is a non-profit body run and supported by academics who have researched and developed these layers, as further detailed in Fig. 1. This Figure illustrates that the three layers are decomposed into eight sub-layers.

The layers and sub-layers are an abstraction that represents and considers the enterprise as a whole [8]. For example, a policy, act, regulation or even a strategy is a part of the business layer, while the application systems and data aspects are a part of the information layer. It also highlights that organisational requirements cut across all the layers, and organisational transformation and innovation draws on the layers too. The Layered Enterprise Architecture Development (LEAD) that Fig. 1 depicts has been embodied by the industry practitioners' enterprise standards body LEADing Practice (www.leadingpractice. com) [7].

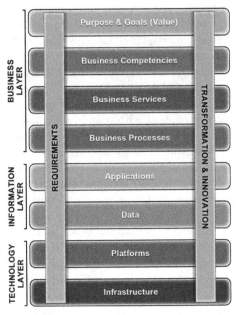

Fig. 1. The layers, and sub-layers

Figure 1 is further dimensioned by Fig. 2, which explicates how the architectural domains (the layers through their sub-layers) are further decomposed into architectural views—i.e. contextual, conceptual, logical and physical. These views are called 'levels' and described shortly. The figure is now a matrix structure, where the layers and sub-layers are the rows, and the levels are the columns.

2.1   An Illustration

To illustrate Fig. 2, Table 1 populates the layer and level structure[1] with metamodel entities. These entities are accordingly referred to as 'meta-objects' which, for simplicity, we shall call objects.

The illustration (which supersedes previous work [4]) will be used to show how we can traverse through the layers and sub-layers, thinking and working simultaneously within and between domains through the decomposition and composition of the objects on all the levels. In so doing, we effortlessly integrate the right concept—i.e. object—across the different sub-layers when interlinking the EA for an organisation [8].

---

[1] The table has the layers in columns not rows, and the levels in rows not columns. This layout allows the table to best fit on the page; it should be the other way round.

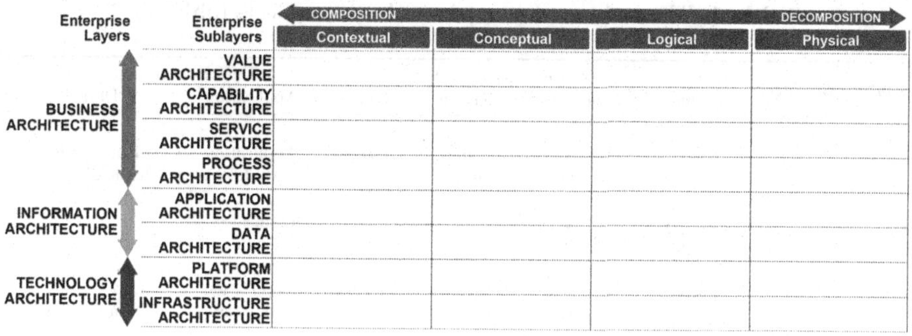

Enterprise Layers	Enterprise Sublayers	COMPOSITION			DECOMPOSITION
		Contextual	Conceptual	Logical	Physical
BUSINESS ARCHITECTURE	VALUE ARCHITECTURE				
	CAPABILITY ARCHITECTURE				
	SERVICE ARCHITECTURE				
	PROCESS ARCHITECTURE				
INFORMATION ARCHITECTURE	APPLICATION ARCHITECTURE				
	DATA ARCHITECTURE				
TECHNOLOGY ARCHITECTURE	PLATFORM ARCHITECTURE				
	INFRASTRUCTURE ARCHITECTURE				

Fig. 2. Layers with levels (contextual, conceptual, logical and physical)

Returning to the table, 'Application Function' is an object under Application in the Information layer at level 2. Likewise, 'Objective' is a business layer Value object at level 3. The table lacks the technology layer but demonstrates the principle of layering, at least for the business and information layers. Also in line with Fig. 2, Table 1's level 1 is the contextual view, level 2 is the conceptual view, level 3 is the logical view, and level 4 is the physical view in architectural view terms.

Many of the objects can exist in the business, information and technology layers. These objects hence can both be repeated or related at more than one level (e.g. vision, mission, strategy, goal, business function, and business service), but are scoped according to their level of abstraction. For example, the strategy object at level 3 reflects an implementation of the strategy set by the highest-level and most abstract contextual depiction of the strategy object at level 1 that in turn is mapped to level 3 through the intermediate conceptual strategy object at level 2. Level 4 is the physical form of the three levels above it. The table shows, for example, a performance indicator so that strategy and other value elements can be measured thereby to determine their effectiveness. Likewise, the most physical form of data sub-layer appears at level 4 (e.g. data table, key/foreign-key/attributes). The other level 4 objects for these and the other sub-layers (competency, service, process and application) can be viewed from the table. A more detailed discussion beyond illustrating the principle as we have described can be found elsewhere [7]. We will, however, explore the objects and their interrelationships as illustrated by Table 1 through Conceptual Structures.

3 Conceptual Structures

Conceptual Graphs (CGs) are a system of logic that express meaning in a form that is logically precise, humanly readable, and computationally tractable. CGs are a conceptual structure that serve as an intermediate language for translating between computer-oriented formalisms and natural languages. CGs graphical representation serve as a readable, but formal design and specification language

Table 1. The meta-model matrix with meta-objects (objects)

Layer & Sub-layer:	Business				Information	
	Value	Competency	Service	Process	Application	Data
Level 1	Vision, Mission	Business Function	Business Service	Business Process	Application Module	Enterprise Data Cluster
	Strategy, Goal	Organizational unit			Organizational unit	
2	Vision, Mission	Business Function	Business Service	Process Step	Application Function	Department Data Cluster
	Strategy, Goal	Organizational unit			Organizational unit	
3	Vision, Mission	Business Function	Business Service	Process Activity	Application Task	Workplace Data Entity
	Strategy, Goal				Transaction Code, System organizational Unit	
						Dimension
	Objective	Business Object		Event	Business Object	
					Data Entity Event	Data Entity
		Business Media / Accounts		Data Object	Data Object (Media)	
		Business Roles	Services Roles	Process Role	Application Roles	
		Business Roles	Service Rules	Process Rules	Application Rules	
4	Performance Indicator	Business Compliance	Service Level Agreement (SLA)	Process Performance Indicator (PPI)	IT Governance	Fact Table Customizing Data Table
						Master Data Table / View
						Transaction Data Table
	Revenue/ CostFlow				System Measurements	Key Foreign Key Describing Attributes

[6]. CGs can thus powerfully represent the formal structure of meta-models while allowing them to be human-readable. A CG (Conceptual Graph) was therefore produced for each sub-layer in Table 1.

Although CGs provide a logical level of rigour, their constituent concepts and relations are essentially put together by hand according to the human's subjective interpretation of the real-world phenomena for it to be captured in a logical structure. This procedure is akin to how the meta-models are produced in practice, using for example Class Diagrams in UML, which CGs can help model [9]. A second form of conceptual structure known as Formal Concept Analysis (FCA) which is used in information science [5]. FCA provides an objective mathematical interpretation of CGs' logical but subjective human interpretations and is brought to bear through the *CGtoFCA* algorithm [1]. A Formal Concept in FCA is the result of when certain conditions are met in a formal *context*:

- A formal context is a triple $\mathbb{K} = (G, M, I)$, where G is a set of objects, M is a set of attributes, and $I \subseteq G \times M$ is a binary (true/false) relation that expresses which objects have which attributes.
- (A, B) is a formal concept precisely when:
  - every object in A has every attribute in B,
  - for every object in G that is not in A, there is some attribute in B that the object does not have,
  - for every attribute in M that is not in B, there is some object in A that does not have that attribute.

The Formal Concepts can then be presented in a lattice, namely a Formal Concept Lattice (FCL), as will be demonstrated shortly.

3.1   The Business Layer

The top layer, the Business Layer, establishes the connections of the enterprise to the environment through the identification of objects that describe the purpose and goal and therefore points both to the source of value and to concerns about the trade-offs necessary to optimise the ability to pursue this value. It further identifies the competencies needed to execute the functions, processes, and services within the environment. These are then used, in conjunction with business functions and other primitives, to organise and aid in the decomposition and organisation of the logical view and physical implementation of the business services and processes. In the following, we will elaborate on the individual sub-layers of the business layer.

Value. The Value architecture sub-layer captures ideas about the vision, mission, strategy policy, act and regulations as well as all the purpose, goal and value that the organisation seeks to create.

The CG (Conceptual Graph) for the Value sub-layer is shown in Fig. 3. It reveals how CGs follow an elementary concept → relation → concept structure that describes the ontology and semantics of the meta-model as explained earlier. The Value CG shows each object name (i.e. Vision, Mission, Strategy, Goal) as a CG type label. To instantiate it as a particular object, a unique identifier appears in the referent field, hence making use of CGs [Type-Label:

Referent] structure. For example, v1V denotes that a object that is Vision (v), Level 1 (1), and V (Value sub-layer). Likewise, g3V for example describes Goal, Level 3, Value and so on. The [Enterprise: @enterprise] concept follows an alternative pattern where @enterprise is a CGs' measure referent [6].

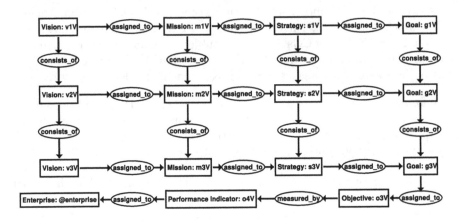

Fig. 3. Value, CGs

The key significance of the [Enterprise: @enterprise] concept is that all the activities that make up an enterprise ultimately point to the enterprise, even though Enterprise is absent in Table 1. This follows EA's holistic perspective. To draw from a building architect's analogy, architecture ranges "From the blank piece of paper to the last nail in the wall." EA follows the same principle, bringing all the objects at all the levels within a sub-layer to the same single point i.e. [Enterprise: @enterprise] being the organisation (that in EA terms is the enterprise, which accounts for all kinds of organisations not just profit-making enterprises).

The relations (e.g. (assigned_to)) describe the interrelationships between the objects in the table. Essentially the (assigned_to) relation refers to a horizontal relation usually in the same sub-layer while (consists_of) is a vertical relation between the levels in the sub-layers. (There is no associated layer, sub-layer or level for Enterprise as it reflects the above-described culmination of all the sub-layers, and—as we shall see—all the levels). The relation (measured-by) has its usual meaning.

Figure 4 shows the FCL (Formal Concept Lattice) for the Value sub-layer. It is the result of the *CGtoFCA* algorithm transforming the object→ relation → object triples in the CG of Fig. 3 to object⌢relation→ object binaries[2]. An

---

[2] The CGs are drawn in CharGer (http://charger.sourceforge.net/) as it has support for the CGIF (CG Interchange Format) in ISO/IEC 24707 Common Logic. The CGIF is passed through *CGtoFCA* then visualised as an FCL in FCA Concept Explorer (http://conexp.sourceforge.net/).

example binary is Vision:v1V⌒assigned_to→Mission: m1V. The neatly displayed lattice shows that [Enterprise: @enterprise] is bottommost i.e. at the *infimum* of the FCL, and highlighted by the bold rectangle in Fig. 4. The topmost formal concept in a FCL is the *supremum*. In this case the supremum is represented by [Vision: v1V], which traverses downwards through the lines (pathways) connecting the intermediate concepts in the Value FCL case culminating in [Enterprise: @enterprise].

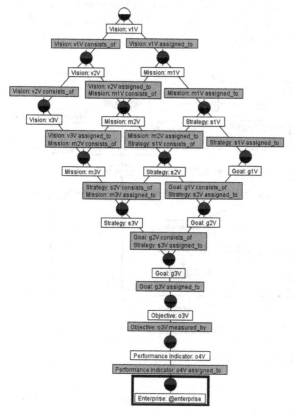

Fig. 4. Value, FCL

The above downward traversal is denoted as that formal concept's *extent*, meaning all the FCA objects below it. As [Enterprise: @enterprise] is at the infimum that means *all* the lattice's objects are in [Vision: v1V]'s extent. This is because in *CGtoFCA* a CG concept becomes an FCA object. Given each CG concept represents a meta-model object in our case, that object in effect becomes an FCA object. Hence why in the Value FCL all the objects are in the extent of [Vision: v1V].

The object⌒relation (e.g. Vision: v1V assigned_to) part of the binary in *CGto FCA* likewise becomes an FCA attribute. The upward (as opposed to downward) traversal is denoted as that formal concept's *intent*, meaning all the FCA attributes (as opposed to objects) that are through the lines (pathways) above it (as opposed to below it). Hence in our case all the object⌒relation attributes going upwards from a formal concept are in that formal concept's intent. If the formal concept is the infimum then that would mean *every* attribute in the FCL. As the infimum in the Value FCL is populated by the [Enterprise: @enterprise] FCA object then all the attributes are in its intent including Vision: v1V assigned_to.

Competency. Figure 5 shows the CG for the Competency architecture sub-layer. Some of the CG concepts in Fig. 5 are shaded to highlight where they appear in the other sub-layers shown by the sub-layer and level Figs. 1 and 2 shown earlier. The referent can be inspected to reveal the other sub-layer (e.g. The '..V' in the referent for `Goal` shows it is in the Business Layer, under the Value sub-layer; the '..A' in the referent for `Business Object` shows it is in the Information Layer, under the Application sub-layer). There is an (`occurrence_copy`) relation too. Essentially, this relation describes two concepts that are similar but are not co-referent (i.e. do not have the same referent), which would make them the same. For example, [`Business Function: bf1C`] → (`occurrence_copy`) → [`Business Service: bs1S`]. The rationale for such a relationship is further detailed elsewhere [4].

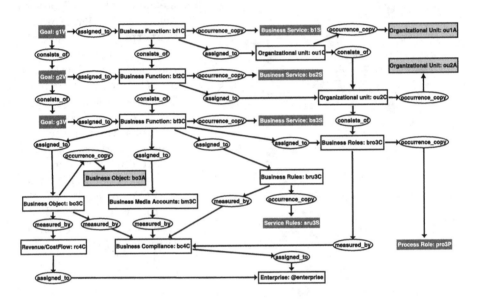

Fig. 5. Competency, CGs

Again the same mapping through *CGtoFCA* is applied and Fig. 6 shows the resulting FCL. This time [`Enterprise: @enterprise`] is not bottom-most, highlighted by the bold rectangles in Fig. 6. The outcome is due to the concepts in the CG, such as [`Business Service: b1S`], [`Service Rules: sru3S`], [`Process Role: pro3P`] that have their identical concept in another business sub-layer (e.g. S for Service, P for Process). These concepts do not have [`Enterprise: @enterprise`] in their extent for the Competency meta-model. Likewise [`Business Object: bo3A`], and [`Organizational Unit: ou1A`] in A the Application sub-layer do not end up at the CG concept [`Enterprise: @enterprise`] unlike the Value CG Fig. 5 above. This again is because the [`Enterprise: @enterprise`] is not in their extent.

The Competency meta-model henceforth has dependencies with the other sub-layers that will only be resolved when the CGs in this sub-layer are joined with the identical, corresponding CG concepts in those other sub-layers. This is

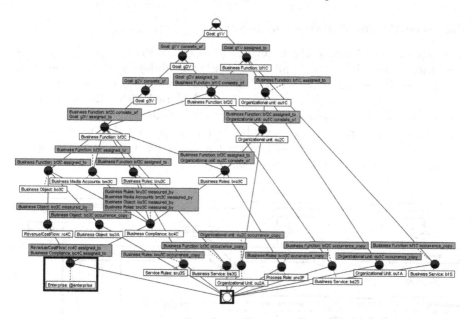

Fig. 6. Competency, FCL

possible through the CGs join operation, where the CG concepts have the same referent i.e. are co-referent [6]. For example the Value sub-layer concept [Goal: g1V] can join with its counterpart in Competency as they share 'g1V'. This operation applies to all matching referents (co-referents) across all the sub-layers. If, when all the sub-layers are thus joined, all the paths lead to [Enterprise: @enterprise] then, together, the (Enterprise) Architectural principle of arriving at that 'last nail in wall—i.e. [Enterprise: @enterprise]—is achieved.

While a simple inspection of the CG for this sub-layer without the FCL reveals the incomplete arrival to [Enterprise: @enterprise], the FCL—which is computer generated rather than hand-drawn—horizontally lays out the objects according to their levels. Where they do not eventually point to [Enterprise: @enterprise], the levels look more uneven. Compare the horizontal layout of the levels in Fig. 6 with Fig. 4 for example. As well as highlighting that the levels look uneven, the concepts in Fig. 5 that do not eventually point to [Enterprise: @enterprise] are further highlighted by being shown at the same bottom level part of the lattice as [Enterprise: @enterprise]. Manual inspection of the CG makes identifying these much harder, especially for the much more comprehensive meta-models encountered in practice than the simple illustration given in Table 1. And we haven't accounted for discovering and rectifying errors in drawing the CGs, such as the arrows in the CGs pointing in the wrong direction or mistyped concepts as simple examples from other work have demonstrated [1,4].

3.2   The Information Systems Layer

The Information Architecture Layer describes the objects, semantic relations and deliverables within the Application and Data sub-layers, and are the main components for both Application Architecture, Data Architecture and Information Architecture. The maps, matrices and models used within the Application and Data sub-layers illustrate how their objects such as data goals, data flows, data services, data requirements and data components are linked to application goals, information flows, information services, application requirements, application flows and applications components.

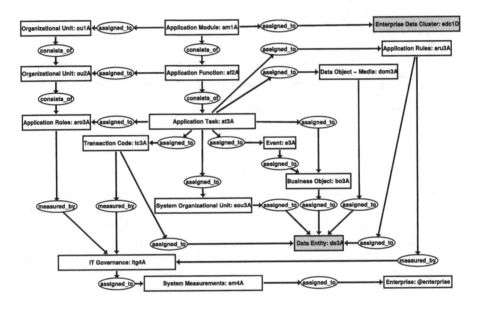

Fig. 7. Application, CGs

Application. Due to space constraints the CG and FCL for the **Service** and **Process** architecture sub-layers are not shown. They have appeared in earlier work [4]. Figure 7 depicts the Application architecture sub-layer CG. Figure 8 evidences that [Enterprise: @enterprise] is not bottommost i.e. not at the infimum. That is because of the object⌢relation attributes that are outside the intent of the level 4 key performance indicator (KPI) object [System Measurements: sm4A], which evaluates the Application sub-layer. Also emerging in the middle of the lattice is another formal concept without its own object. This formal concept appears as a while circle, and is highlighted in Fig. 8 by a bold rectangle. So far such a formal concept has only occurred at the infimum when [Enterprise: @enterprise] is not bottommost, which also happens to be the case in Fig. 8. We can follow the intent and extent from and to the highlighted formal concept to get a sense of what name we might give this object, or confirm that it's simply warranted thus doesn't need its own object. Pertinent to us however is what appears at the infimum.

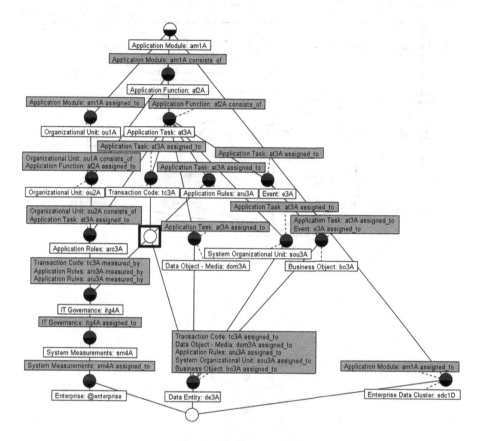

Fig. 8. Application, FCL

Data. Figure 9 depicts the Data architecture sub-layer CG. Figure 10's FCL evidences that [Enterprise: @enterprise] is bottommost i.e. at the infimum. Like Value, the extent of all the attributes from the topmost formal concept i.e. the supremum is [Enterprise: @enterprise] including from all the relevant KPIs (level 4 objects) including [System Measurements: sm4A]. In this sub-layer all its concepts (objects) extend to the enterprise, remembering that it includes objects from other sub-layers (i.e. process, service and application) as highlighted by the shaded CG concepts in Fig. 9. Working back, the dependencies started at the Value architecture sub-layer, and [Vision: v1V] in particular. The meta-models for each sub-layer make up the parts of the whole meta-model for the organisation.

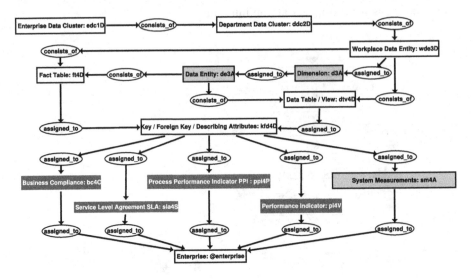

Fig. 9. Data, CGs

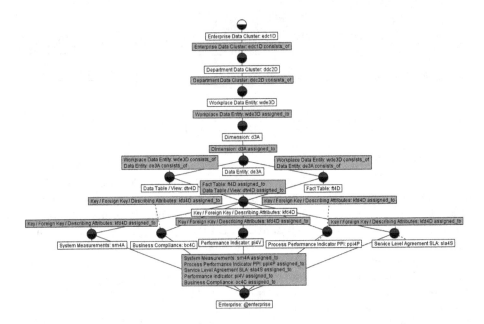

Fig. 10. Data, FCL

4   The Whole Meta-Model

Figure 11 shows the FCL when each CG meta-model for its architecture sub-layer (Value, Competency, Service, Process, Application and Data) are all combined through the co-referent links. Though not shown, it can be appreciated that even for our elementary illustration that would be one, huge, unwieldy CG.

Indeed the FCL Fig. 11 that is generated from this CG looks complex, and the names of the FCA objects and attributes are omitted to avoid cluttering the lattice. (Also though not remarked on here, a number of formal concepts that appear in the middle do not have their own objects, denoted by the clear circles like the circle that appeared for Application.) What is evident nonetheless is that its infimum is a solid circle, highlighted by the bold rectangle in Fig. 11. That is our single point of interest, thus obviating the need to visualise Fig. 11 at all. If its infimum object was shown it would be [Enterprise: @enterprise]. The layered meta-models thereby demonstrate that the organisation's way of thinking and its way of working across all its layers through the sub-layers are aligned.

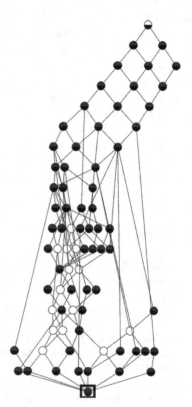

Fig. 11. Combined FCL

Even though the meta-model for each sub-layer apart from Value and Data did not have [Enterprise: @enterprise] as its infimum, when combined into Fig. 11 [Enterprise: @enterprise] became the infimum. It reminds the organisation that it parts (e.g. departments, business, informatics and technical experts) depend on each other, and through the formal concepts explicates through which shared objects that they have to align and communicate through. A good meta-model, as our elementary illustration shows must, in the combined meta-model have an supremum that is the architectural 'blank piece of paper' and an infimum that represents the architectural 'last nail in the wall'. In our illustration that was [Vision: v1V] and [Enterprise: @enterprise] respectively.

5   Conclusions

In EA, thinking and working in architecture domains alone is counterproductive. Through reconsidering the domains as architecture layers and sub-layers, and using levels of contextual, conceptual, logical and physical abstraction within these sub-layers, we have been able to open up the objects' multiple interaction points within and across the various layers, sub-layers and levels that better architect the organisation. From this agile approach, we can think and work with an EA in which the meta-models can repeatedly point to a single truth as opposed to the divergent meta-models that have characterised EA frameworks. Thereby, the organisation is less likely to reinvent the wheel needlessly. We have portrayed how the layered EA can be enhanced by Formal Concepts. The use of CGs (Conceptual Graphs) and FCA (Formal Concept Analysis) through the *CGtoFCA* algorithm provided a formal underpinning to the meta-models, pinpointing the direction of the interdependencies throughout the architecture layers, sub-layers and levels. Through the co-referent links it revealed how meta-models could be aligned towards that single truth and, along the way, without having to generate one large, unwieldy meta-model.

References

1. Andrews, S., Polovina, S.: Exploring, reasoning with and validating directed graphs by applying formal concept analysis to conceptual graphs. In: GKR 2017, Revised Selected Papers. LNAI, vol. 10775, pp. 3–28. Springer, Heidelberg (2018). http://www.springer.com/gb/book/9783319781013
2. The Open Group: 30. content metamodel (2018). http://pubs.opengroup.org/architecture/togaf92-doc/arch/chap30.html
3. Oberle, D.: How ontologies benefit enterprise applications. Semant. Web J. **5**(6), 473–491 (2014)
4. Polovina, S., Scheruhn, H.-J., von Rosing, M.: Modularising the complex meta-models in enterprise systems using conceptual structures. In: Developments and Trends in Intelligent Technologies and Smart Systems, pp. 261–283. IGI Global, Hershey (2018). ID: 189437
5. Priss, U.: Formal concept analysis in information science. Ann. Rev. Info. Sci. Technol. **40**(1), 521–543 (2006)
6. Sowa, J.F.: Conceptual graphs. In: Handbook of Knowledge Representation, Foundations of Artificial Intelligence, vol. 3, pp. 213–237. Elsevier, Amsterdam (2008)
7. von Rosing, M.: Using the business ontology to develop enterprise standards. Int. J. Conceptual Struct. Smart Appl. (IJCSSA) **4**(1), 48–70 (2016). ID: 171391
8. von Rosing, M., Urquhart, B., Zachman, J.A.: Using a business ontology for structuring artefacts: example - northern health. Int. J. Conceptual Struct. Smart Appl. (IJCSSA) **3**(1), 42–85 (2015). ID: 142900
9. Wei, B., Delugach, H.S., Colmenares, E., Stringfellow, C.: A conceptual graphs framework for teaching UML model-based requirements acquisition. In: 2016 IEEE 29th International Conference on Software Engineering Education and Training (CSEET), pp. 71–75, April 2016

Posters

Visualizing Conceptual Structures Using FCA Tools Bundle

Levente Lorand Kis, Christian Săcărea, and Diana-Florina Şotropa[✉]

Babeş-Bolyai University, Cluj-Napoca, Romania
kis_lori@yahoo.com, {csacarea,diana.sotropa}@cs.ubbcluj.ro

Abstract. Formal Concept Analysis (FCA) is a prominent field of applied mathematics organising collections of knowledge - formal concepts - as conceptual landscapes of knowledge. FCA proved to be a promising theory to extract, analyse and visualise conceptual structures arising from various data structures. One of the strengths of FCA is the elegant, intuitive and powerful graphical representation of landscapes of knowledge as concept lattices. The purpose of this paper is to present FCA Tools Bundle and its various features, which is a bundle of tools for dyadic, many-valued, triadic and even polyadic FCA.

1 Introduction

Formal Concept Analysis (FCA) is dealing with collections of knowledge in order to detect, extract, process and represent patterns in various data sets. Following the Conceptual Landscapes of Knowledge paradigm [10], we present FCA Tools Bundle, a collection of tools covering the dyadic case, many-valued contexts, scale building and conceptual browsing. Polyadic data sets can be imported, the corresponding concept sets being calculated by using Answer Set Programming (ASP). In the triadic case, the visualization of the correspondent conceptual structures is based a local navigation paradigm in triadic data sets [8], while for higher-adic concept sets constraint based navigation is implemented [9]. FCA Tools Bundle can also be used to visualize conceptual structures arising from pattern structures [4] by an appropriate scale building. It also offers the possibility to compute analogical proportions in order to mine and represent analogies between formal concepts [6]. Moreover, following the ideas of Wille, we work on a separate feature, navigation in 3D landscapes of knowledge by a 3D visualization of concept lattices using some virtual reality hardware. Since this is work in progress, it has not been included in this short presentation. FCA Tools Bundle is a web based open access collaborative platform where users can share data, create public and private groups or can enter virtual conceptual exploration rooms.

© Springer International Publishing AG, part of Springer Nature 2018
P. Chapman et al. (Eds.): ICCS 2018, LNAI 10872, pp. 193–196, 2018.
https://doi.org/10.1007/978-3-319-91379-7_15

2   Related Work

There is a long list of software tools, developed in the last 30 years in order to support FCA based knowledge visualization.[1] Among them we briefly recall Concept Explorer which supports context processing, clarification and reduction of the context, computing the concept set and the conceptual hierarchy, the ToscanaJ Suite [2] which comprises three components: Elba (create the conceptual scales), Toscana (browse the conceptual schema) and Siena (display the result of conceptual scaling of the entire many-valued context and by that the entire conceptual structure of it). FCABedrock[2] handles many-valued contexts and supports discrete and progressive scaling for continuous attributes.

Most of the algorithms for computing dyadic formal concepts cannot be extended efficiently for the triadic case. R. Jäschke implemented Trias [3], one of the most popular algorithms, which offers the possibility to set the minimum support of the components in the input configuration file. Lattice miner[3] is an FCA software tool for the construction, visualization and manipulation of concept lattices, which allows the generation of formal concepts and association rules. Formal Concept Analysis Research Toolbox[4] (FCART) is a software developed especially for the analysis of unstructured data [7] and is intended for knowledge discovery. LatViz is one of the newest tools developed within the FCA community which introduces interaction with expert, visualization of pattern Structures, AOC posets, concept annotations, filtering concept lattice based on several criteria and an intuitive visualization of implications. By this, the user can effectively perform an interactive exploration over a concept lattice which is a basis for a strong user interaction with WOD for data analysis [1].

3   FCA Tools Bundle - Description and Features

A first description of the FCA Tools Bundle features for dyadic and triadic FCA had been presented in [5]. The key features offered by the tool for a dyadic context are to compute and visualize the object, attribute and concept sets, the incidence relations and the corresponding concept lattice. For triadic contexts, one can compute first the concept list using ASP and then start a local visualization of parts of the triadic context which enables the navigation paradigm based on dyadic projections [8]. For polyadic data sets, one can narrow down the search space by using user defined constraints (on objects, attributes, conditions, states, etc.) and then compute the corresponding resulting concept sets. We proposed an ASP encoding for the membership constraint satisfiability problem and described an interactive search scenario [8]. As far as we know, this is the only software tool allowing navigation in polyadic concept sets and visualization of triconcept sets by using local navigation.

---

[1] An overview of this developing effort is maintained by Uta Priss on her page
http://www.upriss.org.uk/fca/fcasoftware.html.

[2] https://sourceforge.net/projects/fcabedrock/.

[3] https://sourceforge.net/projects/lattice-miner.

[4] https://cs.hse.ru/en/ai/issa/proj_fcart.

ToscanaJ was for many years the only tool to handle many-valued contexts. To overcome the difficulties and drawbacks of ToscanaJ, FCA Tools Bundle offers the possibility to build conceptual scales, either predefined or custom scales and by thus to browse and visualize the conceptual structures of knowledge of a many-valued context. Conceptual scaling is a process of transforming multi-valued contexts into unary-valued ones. A (conceptual) scale is a formal context that determines this procedure for a certain many-valued attribute. In order to create a scale in FCA Tools Bundle the following steps are required:

- **Select a source:** the tool supports two sources types from where one may build scales: database and csv.
- **Provide General Scale Data:** the tool request to fill in the name of the scale, select a table for your scale and then select the type of the scale. Currently the tool supports the nominal, ordinal, interordinal, grid or custom scales.
- **Provide Type Specific Data:** In order to build a nominal scale you need to select the column on which to build the scale. For an ordinal scale you need to define the column on which to build the scale, the order of the scale (increasing or decreasing), the bounds of the scale (include or exclude) and the actual values. For an interordinal scale you need to define the column on which to build the scale, which side includes the bounds and the actual values. For a grid scale you need to define the two columns on which to build the scale, the order for each of the two columns, the bounds for each of the two columns and the values for each of the two columns. In order to build a custom scale you need to create an incidence table defining the custom scale. Custom scales are used for advanced cases where the elementary scale types are not expressive enough.

FCA Tools Bundle is able to compute and display the concepts of a concept lattice which are in analogical complex relation by using the ASP based approach presented in [6]. Analogical complexes are formed by using analogy between four subsets of objects in place of the initial binary relation. They represent subsets of objects and attributes that share a maximal analogical relation. This feature can be very useful since it is interesting to find relations between concepts that are not directly linked in a concept lattice.

4 Conclusions and Future Work

In this paper, we presented FCA Tools Bundle, a platform that offers, for now, features of visualization and navigation for polyadic FCA. We have improved concept lattices generation using a detection collision algorithm, in order to avoid manually arranging the concept lattice for concept visibility. Moreover, we have shown how concept lattices can be used for a triadic navigation paradigm based on appropriately defined dyadic projections. We have implemented analogical proportions between formal concepts and have discussed various features of this tools bundle.

Further developments will include an AI assistant for navigation in large concept lattices, a Temporal Concept Analysis tool, as well as 3D navigation feature by using specific VR hardware.

In conclusion, we believe that the presented version of FCA Tools Bundle brings an important contribution to the collection of FCA tools, by implementing functionalities of visualization and navigation in a large variety of concept sets, which, to the best of our knowledge, are not present in any other tool.

References

1. Alam, M., Le, T.N.N., Napoli, A.: Latviz: a new practical tool for performing interactive exploration over concept lattices. In: Proceedings of the Thirteenth International Conference on Concept Lattices and Their Applications, 18–22 July 2016, Moscow, Russia, pp. 9–20 (2016)
2. Becker, P., Correia, J.H.: The ToscanaJ suite for implementing conceptual information systems. In: Ganter, B., Stumme, G., Wille, R. (eds.) Formal Concept Analysis. LNCS (LNAI), vol. 3626, pp. 324–348. Springer, Heidelberg (2005). https://doi.org/10.1007/11528784_17
3. Jäschke, R., Hotho, A., Schmitz, C., Ganter, B., Stumme, G.: Discovering shared conceptualizations in folksonomies. Web Semant.: Sci. Serv. Agents World Wide Web 6(1), 38–53 (2008)
4. Kaytoue, M., Codocedo, V., Buzmakov, A., Baixeries, J., Kuznetsov, S.O., Napoli, A.: Pattern structures and concept lattices for data mining and knowledge processing. In: Bifet, A., et al. (eds.) ECML PKDD 2015. LNCS (LNAI), vol. 9286, pp. 227–231. Springer, Cham (2015). https://doi.org/10.1007/978-3-319-23461-8_19
5. Kis, L.L., Sacarea, C., Troanca, D.: FCA tools bundle - a tool that enables dyadic and triadic conceptual navigation. In: Proceedings of the 5th International Workshop "What can FCA do for Artificial Intelligence"? Co-located with the European Conference on Artificial Intelligence, FCA4AI@ECAI 2016, The Hague, The Netherlands, 30 August 2016, pp. 42–50 (2016)
6. Miclet, L., Nicolas, J.: From formal concepts to analogical complexes. In: Proceedings of the Twelfth International Conference on Concept Lattices and Their Applications, Clermont-Ferrand, France, 13–16 October 2015, pp. 159–170 (2015)
7. Neznanov, A., Ilvovsky, D.I., Parinov, A.: Advancing FCA workflow in FCART system for knowledge discovery in quantitative data. In: Proceedings of the Second International Conference on Information Technology and Quantitative Management, ITQM 2014, National Research University Higher School of Economics (HSE), 3–5 June 2014, Moscow, Russia, pp. 201–210 (2014)
8. Rudolph, S., Săcărea, C., Troancă, D.: Towards a navigation paradigm for triadic concepts. In: Baixeries, J., Sacarea, C., Ojeda-Aciego, M. (eds.) ICFCA 2015. LNCS (LNAI), vol. 9113, pp. 252–267. Springer, Cham (2015). https://doi.org/10.1007/978-3-319-19545-2_16
9. Troanca, D.: Conceptual visualization and navigation methods for polyadic formal concept analysis. In: Proceedings of the Twenty-Fifth International Joint Conference on Artificial Intelligence, IJCAI 2016, 9–15 July 2016, New York, NY, USA, pp. 4034–4035 (2016)
10. Wille, R.: Conceptual landscapes of knowledge: a pragmatic paradigm for knowledge processing. In: Gaul, W., Locarek-Junge, H. (eds.) Classification in the Information Age. Studies in Classification, Data Analysis, and Knowledge Organization, pp. 344–356. Springer, Heidelberg (1999). https://doi.org/10.1007/978-3-642-60187-3_36

Node-Link Diagrams as Lenses
for Organizational Knowledge Sharing
on a Social Business Platform

Anne-Roos Bakker[1(✉)] and Leonie Bosveld-de Smet[2]

[1] Embrace SBS, Groningen, Netherlands
anne-roos.bakker@embracesbs.com
[2] Department of Information Science, University of Groningen,
Groningen, The Netherlands

Abstract. This poster shows to what extent simple node-link diagrams based on social business network data reflect knowledge sharing behavior within an organization. We show that the number of in- and outgoing messages and key positions of employees in the network are not necessarily good indicators of knowledge sharing. Filtering work-related messages and encoding visually specific attributes of (sets of) employees provide more reliable diagrams that allow us to gain insight into the way social media tools are used for knowledge sharing purposes within an organization. It turns out that currently, most organizations do not exploit social business software to its full knowledge sharing potential.

Keywords: Social networks · Node-link diagrams
Knowledge sharing · Enterprise microblogging

1 Introduction

Social business software introduces social media tools, such as a microblogging service, in a corporate context [3]. It intends to facilitate knowledge sharing, collaboration and connection that go beyond hierarchical, inter- and intradepartmental, and geographical frontiers within a company or organization. It allows to create networks of groups of employees with complementary knowledge, common interests, and to make information streams more directed. Moreover, it may stimulate social involvement, which is important for connecting employees and creating a feeling of community [2]. For knowledge sharing it is crucial that employees with expert knowledge are active on the platform, when their expertise is needed by other employees, departments or locations. These employees are the so-called knowledge holders. Also it is important that employees in higher levels of the organizational hierarchy, such as managers and CEO's, state their views on the platform [1]. Their presence shows the value to use social business tools. Social business software also allows non-collocated (sets of) individuals to share knowledge. A common observation is that most online activity depends

© Springer International Publishing AG, part of Springer Nature 2018
P. Chapman et al. (Eds.): ICCS 2018, LNAI 10872, pp. 197–200, 2018.
https://doi.org/10.1007/978-3-319-91379-7_16

on knowing each other offline. Social media tools in a corporate context may help to motivate knowledge sharing across departments and geographical locations. When adopted by organizations, social business software generates data that may reflect the extent to which employees share knowledge, collaborate, and connect by using the tools afforded. These data are worth studying in order to get insight into the question whether and how the intended goals of social business software tools are achieved. The tools we propose to use to evaluate the current knowledge sharing activity within organizations, as revealed by social business software data, are simple node-link diagrams, easily generated by visualization tools such as NodeXL and Gephi.

2   Method

To visualize the social network behavior of employees in social business software, we use simple node-link diagrams. Nodes represent different entities. Nodes correspond to individual persons, or professions related to them, or sets of persons, working in different departments and at different geographical locations. The relationships between the nodes are represented by edges. The edges are directed, indicating who replies to whom. The edges are weighted, based on the number of posts of one node to another. We have assessed the position of nodes in the diagram by characterizing them as bridges, central nodes and isolated nodes by visual inspection. The node positions in the diagram are determined by the common metrics used in visualization tools. Bridges and central nodes are seen as key nodes with respect to knowledge sharing.

There are several potential indicators within a node-link diagram for knowledge sharing. We have investigated which ones are truthful. First we have combined visual inspection of node-link diagrams with content analysis of in- and outgoing posts. Second we have used the insights gained to improve the diagrams so that they provide more truthful information about knowledge sharing. Due to privacy laws, the figures given in this paper are not based on real customer data.

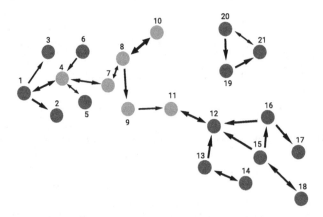

Fig. 1. Managers (blue) and non-managers (red). (Color figure online)

However, they correspond to diagrams generated by social network customer data of different organizations.

3   Towards More Insightful Node-Link Diagrams

Edge weights indicate the intensity of connections between individuals or groups of individuals. As nodes can represent both individuals and groups (departments or locations), weights of edges and key positions can be a result of the amount of posts sent or received by a large group of individuals. Visual enhancement can help to correctly interpret nodes that represent different entities. To add information about the type of entity represented by a node, size can be used to encode the cardinality of the set of persons involved. Edge weights can be interpreted in relation to the size of the node.

In some node-link diagrams, central and bridging roles can be caused by non work-related events. Some users get key positions in the diagram because they send out non work-related posts, which get a lot of replies of other employees who show social involvement. Messages that usually get a lot of replies are messages about birth of a child, illness or birthdays. These key positions cannot be associated to work-related knowledge sharing. It is important to analyze post content (by manual and/or machine learning coding) to get insight into the exact link between the post and direct work-related knowledge sharing. Taking into account only specific content allows to get more reliable diagrams.

To get a good overview and interpret correctly what the role of specific individuals is in the network, you can encode different professions by color. To

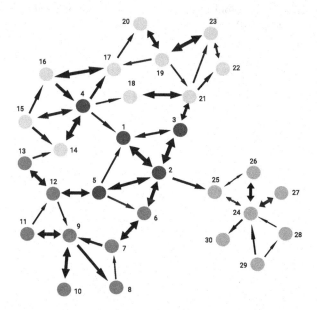

Fig. 2. Separate departments in organizations (different colors). (Color figure online)

get an even more reliable result, you can only select data of messages that contain work-related content. When knowledge sharing is performed in an optimal way, knowledge holders will be visible in key positions. Knowledge seekers know whom to connect with to get an expert answer to their questions.

Platform activity of managers and CEO's are important for the willingness of employees to participate. As a result you would like to see managers in key positions in the network. Figure 1 shows an example of a node-link diagram that positions managers in networks based on observations in current organizations. Managers are represented by blue nodes, other employees by red nodes. It shows that managers connect two different groups, so they are bridges. However, they also send messages to each other or get them from each other.

Especially in large, hierarchical organizations employees do their jobs in separate departments. Figure 2 is an example of a diagram that involve intra-departmental connections. The nodes of employees working in the same department, have the same color. We see that the departments are interconnected, but that most individuals make barely connections with other departments. Especially the green nodes form an isolated cluster.

References

1. Chin, C.P.-Y., Evans, N., Choo, K.K.-R., Tan, F.B.: What influences employees to use enterprise social networks? A socio-technical perspective. In: PACIS, p. 54 (2015)
2. Mäntymäki, M., Riemer, K.: Enterprise social networking: a knowledge management perspective. Int. J. Inf. Manag. **36**(6), 1042–1052 (2016)
3. Riemer, K., Richter, A., Bohringer, M.: Enterprise microblogging. Bus. Inf. Syst. Eng. **2**(6), 391–394 (2010)

Author Index

Agon, Carlos   105
Andreatta, Moreno   105
Andrews, Simon   137
Atif, Jamal   105

Bakker, Anne-Roos   197
Bisquert, Pierre   73
Bloch, Isabelle   105
Bosveld-de Smet, Leonie   197
Braud, Agnès   152
Braun, Tanya   39, 55

Croitoru, Madalina   73

Endres, Dominik   88

Gehrke, Marcel   55

Hanika, Tom   120
Hitzler, Pascal   3

Ibrahim, Mohamed-Hamza   24

Kis, Levente Lorand   193

Le Ber, Florence   152

Mascarade, Pierre   105
McLeod, Kenneth   137
Missaoui, Rokia   24
Möller, Ralf   39, 55
Molnar, Andrea Eva   167

Nica, Cristina   152

Polovina, Simon   176
Priss, Uta   96

Săcărea, Christian   9, 167, 193
Schubert, Moritz   88
Shimizu, Cogan   3
Şotropa, Diana   9
Şotropa, Diana-Florina   193

Thomopoulos, Rallou   73
Troancă, Diana   9

Varga, Viorica   167
von Rosing, Mark   176

Yun, Bruno   73

Zumbrägel, Jens   120

Printed in the United States
By Bookmasters

Printed in the United States
By Bookmasters